# Crossing the Chasm

# CROSSING THE CHASM

Marketing and Selling Technology
Products to Mainstream Customers

## Geoffrey A. Moore

with a foreword by
Regis McKenna

HarperBusiness
*A Division of HarperCollinsPublishers*

Library of Congress Cataloging-in-Publication Data

Moore, Geoffrey A., 1946-
    Crossing the chasm : marketing and selling technology products to main-
stream customers / Geoffrey A. Moore : with a foreword by Regis McKanna.
      p.    cm.
    Includes index.
    ISBN 0-88730-519-9
    1. Selling—technology.   2. High technology industries.   3. Marketing.
I. Title.
HF5439.T42M66     1991                           91-24260
658.8—dc20                                      CIP

Printed in the United States of America

94 95 96 SWD/HC 15 14 13 12

To
Marie

# Contents

# Foreword

Within an ever-changing society, marketing represents the ongoing effort to keep the means of production—our products and services—in touch with evolving social and personal conditions. That "keeping in touch" has become our greatest challenge.

In an era when the pace of change was slower, the variety of products and services fewer, the channels of communication and distribution less pervasive, and the consumer less sophisticated, marketing could enjoy prolonged periods of relative stability, reaping profits from "holding the customer constant" and optimizing the other variables. That is no longer the case.

We live in an age of choice. We are continually bombarded with purchasing alternatives in every aspect of our lives. This in turn has led us to develop an increasingly sophisticated set of defenses, so that any company seeking to establish a "brand loyalty" in us is going to be hard-pressed to succeed. We demand more and more from our purchases and our suppliers, leading to increasingly fragmented markets served by products that can be customized by design, programmability, service, or variety.

There is a wonderful analogy to all this in the world of high technology. Behind the astounding proliferation of electronic systems, infiltrating our entertainment centers, our phones, our cars, and our kitchens, lies a technology called *application-specific integrated circuits*, or ASICs. These are tiny microprocessors that are producible in high volume up to the last layer, which is then designed *by the customers* to add the final veneer of personality needed for their specific product. ASICs embody many of the fundamental elements of modern marketing—radical customizability overlaid onto a constant and reliable foundation, dramatically shortened times to market, relatively small production runs, and an intense focus on customer service. They exemplify the remaking of our means of production to accommodate our changing social and personal needs.

As uplifting as all of this sounds in theory, in practice it represents a great challenge not only to our economic institutions but to the human spirit itself. We may celebrate change and growth, but that does not make either one the less demanding or painful. Our emerging and evolving markets are demanding continual adaptation and renewal, not only in times of difficulty but on the heels of our greatest successes as well. Which of us would not prefer a little more time to savor that success, to reap a little longer what we cannot help but feel are our just rewards? It is only natural to cling to the past when the past represents so much of what we have strived to achieve.

This is key to *Crossing the Chasm*. The chasm represents the gulf between two distinct marketplaces for technology products—the first, an early market dominated by early adopters and insiders who are quick to appreciate the nature and benefits of the new development, and the second a mainstream market representing "the rest of us," people who want the benefits of new technology but who do not want to "experience" it in all its gory details. The transition between these two markets is anything but smooth.

Indeed, what Geoff Moore has brought into focus is that, at the time when one has just achieved great initial success in launching a new technology product, creating what he calls early market wins, one must undertake an immense effort and radical transformation to make the transition into serving the mainstream market. This transition involves sloughing off familiar entrepreneurial marketing habits and taking up new ones that at first feel strangely counterintuitive. It is a demanding time at best, and I will leave the diagnosis of its ailments and the prescription of its remedies to the insightful chapters that follow.

If we step back from this chasm problem, we can see it as an instance of the larger problem of how the marketplace can cope with change in general. For both the customer and the vendor, continually changing products and services challenge their institution's ability to absorb and make use of the new elements. What can marketing do to buffer these shocks?

Fundamentally, marketing must refocus away from *selling product* and toward *creating relationship*. Relationship buffers the shock of change. To be sure, the specific product or service provided remains the fundamental basis for economic exchange, but it must not be treated as the main event. There is simply too much change in this domain for anyone to tolerate over the long haul. Instead, we must direct our attention toward creating and maintaining an ongoing customer relationship, so that as things change and stir in our immediate field of activity, we can look up over the smoke and dust and see an abiding partner, willing to cooperate and adjust with us as we take on our day-to-day challenges. Marketing's first deliverable is that partnership.

This is what we mean when we talk about "owning a market." Customers do not like to be "owned," if that implies lack of choice or freedom. The open systems movement in high tech is a clear example of that. But they do like to be "owned" if what that means is a vendor taking ongoing responsibility for the success of their joint ventures. Ownership in this sense means abiding commitment and a strong sense of mutuality in the development of the marketplace. When customers encounter this kind of ownership, they tend to become fanatically loyal to their supplier, which in turn builds a stable economic base for profitability and growth.

How can marketing foster such relationships? That question has driven the development of Regis McKenna Inc. since its inception. We began in the 1970s in our work with Intel and Apple where we tried to set a new tone around the adoption of technology products, to capture the imagination of a marketplace whose attentions were directed elsewhere. Working with Intel, Apple, Genentech and many other new technology companies, it became clear that traditional marketing approaches would not work. Business schools in America were educating their students to the ways of consumer marketing, and these graduates assumed that marketing was generic. Advertising and brand awareness became synonymous with marketing.

In the 1980s intense competition, even within small niches,

created a new environment. With everyone competing for the customer's attention, the customer became king and demanded more substance than image. Advertising, as a medium of communication, could not sustain the kind of relationship that was needed for ongoing success. Two reasons in particular stood out. First, as Vance Packard, in *The Hidden Persuaders*, and others educated the American populace to the manipulativeness of advertising, its credibility as a means of communication deteriorated. This was an extremely serious loss when it came to high-tech purchase decisions, because of what IBM used to call the "FUD factor"—the fear, uncertainty, and doubt that can plague decision makers when confronted with such an unfamiliar set of products and services. Just when they most want to trust in the communication process, they are confronted with an ad that they believe may be leading them astray.

The second problem with advertising is that it is a one-way mechanism of communication. As the emphasis shifts more and more from selling product to creating relationship, the demand for a two-way means of communication increases. Companies do not get it right the first time. To pick two current market-leading examples, the first Macintosh and the first release of Windows simply were not right—both needed major overhauls before they could become the runaway successes they represent today. This was only possible by Apple and Microsoft keeping in close touch with their customers and the other participants that make up the PC marketplace.

The standard we tried to set at RMI was one of education not promotion, the goal being to communicate rather than to manipulate, the mechanism being dialogue, not monologue. The fundamental requirement for the ongoing, interoperability needed to sustain high tech is accurate and honest exchange of information. Your partners need it, your distribution channel needs it and must support it, and your customers demand it. People in the 1990s simply will not put up with noncredible channels of communication. They will take their business elsewhere.

At RMI we call the building of market relationships *market relations*. The fundamental basis of market relations is to build and manage relationships with all the members that make up a high-tech marketplace, not just the most visible ones. In particular, it means setting up formal and informal communications not only with customers, press, and analysts but also with hardware and software partners, distributors, dealers, VARs, systems integrators, user groups, vertically oriented industry organizations, universities, standards bodies, and international partners. It means improving not only your external communications but also your internal exchange of information among the sales force, the product managers, strategic planners, customer service and support, engineering, manufacturing, and finance.

To facilitate such relationships implies a whole new kind of expertise from a consulting organization. In addition to maintaining its communications disciplines, it must also provide experienced counsel and leadership in making fundamental marketing decisions. Market entry, market segmentation, competitive analysis, positioning, distribution, pricing—all these are issues with which a successful marketing effort must come to grips. And so we again remade ourselves, adding to market relations a second practice—high-tech marketing consulting.

Today, our practices of marketing consulting and market relations together are tackling the fundamental challenge of the 1990s—helping multiple players in the marketplace build what we call "whole product" solutions to market needs. Whole products represent completely configured solutions. Today, unlike the early 1980s, no single vendor, not even an IBM, can unilaterally provide the whole products needed. A new level of cooperation and communication must be defined and implemented so that companies—not just products—can "interoperate" to create these solutions.

*Crossing the Chasm* reflects much of this emphasis. Moore is a senior member of the RMI staff and has become an integral contributor to the development of our practice. An ex-professor and teacher by trade, he does not shrink from taking the stage to evangelize a new agenda. Part of that agenda is to make original contributions to the marketing discipline, and as you will see in the coming chapters, Geoff has done just that. At the same time, as he himself is quick to acknowledge, his colleagues and his clients have made immense contributions as well, and he is to be commended for his efforts in integrating these components into this work.

Finally, I would just like to say that this work is going to make you think. And the best way to prepare yourself for the fast-paced, ever-changing competitive world of marketing is to prepare yourself to think. This book adds the dimension of creative thinking as a prelude to action. It will change the way you think about marketing. It will change the way you think about market relationships.

Regis McKenna

# Acknowledgments

The book that follows represents two years of writing. It also represents my last 13 years of employment in one or another segment of high-tech sales and marketing. And most importantly, it reflects the last four years I have spent as a consultant at Regis McKenna Inc. During this period I have worked with scores of colleagues, sat in on innumerable client meetings, and dealt with myriad marketing problems. These are the "stuff" out of which this book has come.

Prior to entering the world of high tech, I was an English professor. One of the things I learned during this more scholarly period in my life was the importance of evidence and the necessity to document its sources. It chagrins me to have to say, therefore, that there are no documented sources of evidence anywhere in the book that follows. Although I routinely cite numerous examples, I have no studies to back them up, no corroborating witnesses, nothing.

I mention this because I believe it is fundamental to the way in which lessons are transmitted in universities and the way they are transmitted in the workplace. All of the information I use in day-to-day consulting comes to me by way of word of mouth. The fundamental research process for any given subject is to "ask around." There are rarely any real facts to deal with—not regarding the really important issues, anyway. Some of the information may come from reading, but since the sources quoted in the articles are the same as those one talks to, there is no reason to believe that the printed word has any more credibility than the spoken one. There is, in other words, no hope of a definitive answer. One is committed instead to an ongoing pro-

xv

cess of update and revision, always in search of the explanation that gives the best fit.

Given that kind of world, the single most important variable becomes who you talk with. The greatest pleasure of my past four years at RMI has been the quality of people I have encountered as my colleagues and my clients. In the next few paragraphs I want to acknowledge some of them specifically by name, but I know that by so doing I am bound to commit more than one sin of omission. From those who are not mentioned but who should have been, I ask forgiveness in advance.

Several of my current colleagues have offered ongoing input and criticism of this effort in its various conversational and manuscript forms. These include Paul Hodges, Randy Nickel, Elizabeth Chaney, Ellen Hipschman, Rosemary Remacle, Page Alloo, Karen Kang, Karen Lippe, Greg Ruff, Chris Halliwell, Patty Burke, Joan Naidish, Sharon Colby, and Patrick Corman.

Other colleagues who have since moved on to other ventures also provided wisdom, examples, and support. These include Jennifer Jones, Lee James, Lynn Amato, Bob Pearson, Mary Jane Reiter, Nancy Blake, Wendy Grubow, Jean Murphy, John Fess, Kathy Lockton, Andy Rothman, Rick Redding, Jennifer Little, and Wink Grellis. Then there is that one colleague who has cheerfully provided her hard labor in the copying, mailing, faxing, phoning, coordinating and all else that goes into getting a book out. Thank you, Brete Wirth.

Clients and friends—not mutually exclusive groups, I am happy to say—have also been extremely supportive of this effort, both in critiquing drafts of the manuscript and in contributing to the ideas and examples. In this regard, I would especially like to acknowledge John Rizzo, Sam Darcie, David Taylor, Brett Bullington, Tom Quinn, Tom Loeb, Phil Vertin, Mike Whitfield, Bill Leavy, Ed Sterbenc, Bob Jolls, Bob Healy, Paul Wiefels, Mark and Chuck Dehner, Doug Edwards, Corinne Smith, John Zeisler, Jane Gaynor, Bob Lefkowits, Camillo Wilson, Ed Sattizahn, Jon Rant, John Oxaal, Isadore Katz, and Tony Zingale.

From the hoard of interesting remarks of independent consultants and occasional competitors, many of whom are also good friends, I have pillaged cheerfully whenever I could. These include Roberta Graves, Tony Morris, Sy Merrin, Kathy

Lane, Leigh Marriner, Dick Shaffer, Esther Dyson, Jeff Tarter, and Stewart Alsop.

Then we come to that core group of friends whose importance goes beyond specific contributions to this or that idea or chapter and lodges instead somewhere near support of the soul. These exceptionally special folk include Doug Molitor, Glenn Helton, Peter Schireson, Skye Hallberg, and Steve Flint.

Beyond that, there are three more people without whom this book would not be possible. The first of these is Regis McKenna, my boss, founder of my company and funder of my livelihood, and in many senses the inventor of the high-tech marketing practice I am now trying to extend. The second is Jim Levine, my literary agent, the man who took a look at 200-odd pages of manuscript a year or so ago and allowed as how, although it wasn't a book, it might have possibilities. And the third is Virginia Smith, my editor, who has been guiding me this past year through the bizarre intricacies of the book publishing business.

There remains one last group of people to name, those who have been at the center of almost anything I have ever undertaken: my parents, George and Patty; my brother, Peter; my children, Margaret, Michael, and Anna; and my wife, Marie. I am particularly indebted to Marie, for many reasons that go well beyond this book, but specifically in this instance for making the countless sacrifices and giving the kind of emotional and practical day-to-day support that make writing a book possible, and for being the kind of person that inspires me to undertake such challenges.

# PART I
## DISCOVERING THE CHASM

---

# Introduction

## If Bill Gates Can Be a Billionaire

There is a line from a song in the musical *A Chorus Line*: "If Troy Donahue can be a movie star, then I can be a movie star." Every year one imagines hearing a version of this line reprised in high-tech start-ups across the country: "If Bill Gates can be a billionaire . . ." For indeed, the great thing about high tech is that, despite numerous disappointments, it still holds out the siren's lure of a legitimate get-rich-quick opportunity.

But let us set our sights a little more modestly. Let us say, "If two guys, each named Mike Brown (one from Portland, Oregon, and the other from Lenexa, Kansas), can in 10 years found two companies no one has ever heard of (Central Point Software and Innovative Software), and bring to market two software products that have hardly become household names (PC Tools Deluxe and Smartware) and still be able to cash out in seven figures, then, by God, we should be able to too."

This is the great lure. And yet, as even the Bible has warned us, while many are called, few are chosen. Every year millions of dollars—not to mention countless work hours of our nation's best technical talent—are lost in failed attempts to join this kingdom of the elect. And oh what wailing then, what gnashing of teeth!

"Why me?" cries out the unsuccessful entrepreneur. Or rather, "Why <u>not</u> me?" "Why not us?" chorus his equally unsuccessful investors. "Look at our product. Is it not as good—nay, better—than the product that beat us out? How can you say that Oracle is better than Ingres, Lotus 1-2-3 is better than SuperCalc Five, Macintosh is better than Amiga, or the Intel 80386 is better than the National 32032?" How, indeed? For in fact, feature for feature, the less successful product is often arguably superior.

Not content to slink off the stage without some revenge, this sullen and resentful crew casts about among themselves to find a scapegoat, and whom do they light upon? With unfailing consistency and unerring accuracy, all fingers point to—*the vice-president of marketing*. It is marketing's fault! Oracle outmarketed Ingres, Lotus outmarketed Computer Associates, Apple outmarketed Commodore, Intel outmarketed National. Now we too have been outmarketed. Firing is too good for this monster. Hang him!

While this sort of thing takes its toll on the marketing profession, there is more at stake in these failures than a bumpy executive career path. When a high-tech venture fails, everyone goes down with the ship—not only the investors but also the engineers, the manufacturers, the president, and the receptionist. All those extra hours worked in hopes of cashing in on an equity option—all gone.

Worse still, because there is no clear reason why one venture succeeds and the next one fails, the sources of capital to fund new products and companies become increasingly wary of investing. Interest rates go up, and the willingness to entertain venture risks goes down. Wall Street has long been at wit's end when it comes to high-tech stocks. Despite the efforts of some of its best analysts, these stocks are traditionally undervalued, and exceedingly volatile. It is not uncommon for a high-tech company to announce even a modest shortfall in its quarterly projections and incur a 20 to 30 percent devaluation in stock price on the following day of trading.

There is an even more serious ramification. High-tech inventiveness and marketing expertise are two cornerstones of the

U.S. strategy for global competitiveness. We ceded the manufacturing advantage to other countries long ago. If we cannot at least learn to predictably and successfully bring high-tech products to market, our counterattack will falter, placing our entire standard of living in jeopardy.

With so much at stake, the erratic results of high-tech marketing are particularly frustrating, especially in a society where other forms of marketing appear to be so well under control. Elsewhere—in cars or TVs or microwaves—we may see ourselves being outmanufactured, but not outmarketed. Indeed, even after we have lost an entire category of goods to offshore competition, we remain the experts in marketing these goods to U.S. consumers. Why haven't we been able to apply these same skills to high tech? And what is it going to take for us to finally get it right?

It is the purpose of this book to answer these two questions in considerable detail. But the short answer is as follows: Our current model for how to develop a high-tech market is almost—but not quite—right. As a result, our marketing ventures, despite normally promising starts, drift off course in puzzling ways, eventually causing unexpected and unnerving gaps in sales revenues, and sooner or later leading management to undertake some desperate remedy. Occasionally these remedies work out, and the result is a high-tech marketing success. (Of course, when these are written up in retrospect, what was learned in hindsight is not infrequently portrayed as foresight, with the result that no one sees how perilously close to the edge the enterprise veered.) More often, however, the remedies either flat-out fail, and a product or a company goes belly up, or they progress after a fashion to some kind of limp but breathing half-life, in which the company has long since abandoned its dreams of success and contents itself with once again making payroll.

None of this is necessary. We have enough high-tech marketing history now to see where our model has gone wrong and how to fix it. To be specific, the point of greatest peril in the development of a high-tech market lies in making the transition from an *early market* dominated by a few *visionary* customers to a *mainstream market* dominated by a large block of customers who are predominantly *pragmatists* in orientation. The gap between these two markets, heretofore ignored, is in fact so significant as to warrant being called a *chasm*, and crossing this chasm must be the primary focus of any long-term high-tech marketing plan. A successful crossing is how high-tech fortunes are made; failure in the attempt is how they are lost.

As a consulting principal at Regis McKenna Inc., I, along with my colleagues, have watched countless companies struggle to maintain their footing during this difficult period. It is an extremely difficult transition for reasons that will be summarized in the opening chapters of the book. The good news is that there are reliable guiding principles. The material in this book was born of hundreds of consulting engagements focused on bringing products and companies into profitable and sustainable mainstream markets. The models presented here have been tested again and again and have been found effective. The chasm, in sum, can be crossed.

Like a hermit crab that has outgrown its shell, the company crossing the chasm must scurry to find its new home. Until it does, it will be prey to all kinds of predators. This urgency means that everyone in the company—not just the marketing and sales people—must focus all their efforts on this one end until it is accomplished. Chapters 3 through 7 set forth the principles necessary to guide high-tech ventures during this period of great risk. This section focuses on marketing, because that is where the leadership must come from, but I ultimately argue in the Conclusion that leaving the chasm behind requires significant changes throughout the high-tech enterprise. The book closes, therefore, with a call for new strategies in the areas of finance, organizational development, and R&D.

This book is unabashedly about and for marketing within high-tech enterprises. But high tech can be viewed as a microcosm of larger industrial trends. In particular, the relationship between an early market and a mainstream market is not unlike the relationship between a fad and a trend. Marketing has long known how to exploit fads and how to develop trends. The problem, since these techniques are antithetical to each other, is that you need to decide which one—fad or trend—you are dealing with before you start. It would be much better if you could start with a fad, exploit it for all it was worth, and then turn it into a trend.

That may seem like a miracle, but that is in essence what high-tech marketing is all about. Every truly innovative high-tech product starts out as a fad—something with no known market value or purpose but with "great properties" that generate a lot of enthusiasm within an "in crowd." That's the early market. Then comes a period during which the rest of the world watches to see if anything can be made of this; that is the chasm. If in fact something does come out of it—if a value proposition is discovered that can predictably be delivered to a targetable set of customers at a reasonable price—then a new mainstream market forms, typically with a rapidity that allows its initial leaders to become very, very successful.

The key in all this is crossing the chasm—making that mainstream market emerge. This is a do-or-die proposition for high-tech enterprises; hence, it is logical that they be the crucible in which "chasm theory" is formed. But the principles can be generalized to other forms of marketing, so for the general reader who can bear with all the high-tech examples in this book, useful lessons may be learned.

One of the most important lessons about crossing the chasm is that the task ultimately requires achieving an unusual degree of company unity during the crossing period. This is a time when one should forgo the quest for eccentric marketing genius, in favor of achieving an informed consensus among mere mortals. It is a time not for dashing and expensive gestures but rather for careful plans and cautiously rationed resources—a time not to gamble all on some brilliant coup but rather to focus everyone on making as few mistakes as possible.

One of the functions of this book, therefore—and perhaps its most important one—is to open up the logic of marketing decision making during this period so that everyone on the management team can participate in the marketing process. If prudence rather than brilliance is to be our guiding principle, then many heads are better than one. If marketing is going to be the driving force—and most organizations insist this is their goal—then its principles must be accessible to all the players, and not, as is sometimes the case, be reserved to an elect few who have managed to penetrate its mysteries.

*Crossing the Chasm*, therefore, is written for the entire high-tech community—for everyone who is a stakeholder in the venture, engineers as well as marketeers, and financiers as well. All must come to a common accord if the chasm is to be safely negotiated. And with that thought in mind, let us turn to Chapter 1.

# 1

# High-Tech Marketing Illusion

---

As the final draft of this book is being written, it is the beginning of the 1990s, and once again we are beginning to hear talk about the electric car. Maybe this time it is serious—who is to say? Let's assume that General Motors makes one, and Ford and Chrysler follow. Let's assume the cars work like any other, except they are quieter and better for the environment. Now the question is: When are you going to buy one?

## The Technology Adoption Life Cycle

Your answer to the preceding question will tell a lot about how you relate to the *Technology Adoption Life Cycle,* a model for understanding the acceptance of new products. If your answer is, "Not until hell freezes over," you are probably a very late adopter of technology, what we call in the model a *laggard.* If your answer is, "When I have seen electric cars prove them-selves and when there are enough service stations on the road," you might be a middle-of-the-road adopter, or in the model, the *early majority*. If you say, "Not until most people have made the switch and it becomes really inconvenient to drive a gasoline car," you are probably more of a follower, a member of the *late majority*. If, on the other hand, you want to be the first one on your block with an electric car, you are apt to be an *innovator* or an *early adopter*.

In a moment we are going to take a look at these labels in greater detail, but first we need to understand their significance. It turns out our attitude toward technology adoption becomes significant—at least in a marketing sense—any time we are introduced to products that require us to change our current mode of behavior or to modify other products and ser-vices we rely on. In academic terms, such change-sensitive products are called *discontinuous innovations*. The contrasting term, *continuous innovations*, refers to the normal upgrading of products that does not require us to change behavior.

For example, when Cheer promises you even brighter brights, that is a continuous innovation. You still are washing the same clothes in the same way with the same washing machine. When Ford's new car promises better mileage, when Apple's latest computer promises faster processing times and more storage space, or when Sony promises sharper and brighter TV pictures, these are all continuous innovations. As a consumer, you don't have to change your ways in order to take advantage of these improvements.

On the other hand, if the Sony were a high-definition TV, it would be incompatible with today's broadcasting standards, which would require you to seek out special sources of pro-gramming. This would be a discontinuous innovation because you would have to change your normal TV-viewing behavior. Similarly, if the new Apple computer were based on a RISC microprocessor (like IBM's new RS/6000, for example), it would be incompatible with today's software base. Again, you would be required to seek out a whole new set of software, thereby classifying this too as a discontinuous innovation. Or if the new Ford car, as we just noted, required electricity instead of gasoline, or if the new detergent were a dry-cleaning agent that did not use water, then once again you would have a product incompatible with your current infrastructure of supporting components. In all these cases, the innovation demands significant changes by not only the consumer but also the infrastructure. That is how and why such innovations come to be called discontinuous.

Between *continuous* and *discontinuous* lies a spectrum of demands for change. TV dinners, unlike microwave dinners, did not require the purchase of a new oven, but they did require the purchase of more freezer space. Color-TV programming did not, like VCRs, require investing in and mastering a new technology, but they did require buying a new TV and learning more about tuning and antennas than many of us wanted to learn. The special washing instructions for certain fabrics, the special street lanes reserved for bicycle riders, the special dialing instructions for calling overseas—all represent some new level of demand on the consumer to absorb a change in behavior. Sooner or later, all businesses must make these demands. And so it is that all businesses can profit by lessons from high-tech industries.

Whereas other industries introduce discontinuous innovations only occasionally and with much trepidation, high-tech enterprises do so routinely and as confidently as a born-again Christian holding four aces. From their inception, therefore, high-tech industries needed a marketing model that coped effectively with this type of product introduction. Thus the Technology Adoption Life Cycle became central to the entire sector's approach to marketing. (People are usually amused to learn that the original research that gave rise to this model was done on the adoption of new strains of seed potatoes among American farmers. Despite these agrarian roots, however, the model has thoroughly transplanted itself into the soil of Silicon Valley.)

The model describes the market penetration of any new technology product in terms of a progression in the types of consumers it attracts throughout its useful life:

## Technology Adoption Life Cycle

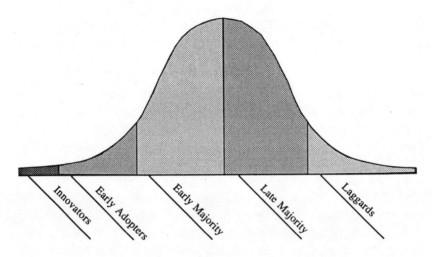

As you can see, we have a bell curve. The divisions in the curve are roughly equivalent to where standard deviations would fall. That is, the early majority and the late majority fall within one standard deviation of the mean, the early adopters and the laggards within two, and way out there, at the very onset of a new technology, about three standard deviations from the norm, are the innovators.

The groups are distinguished from each other by their characteristic response to a discontinuous innovation based on a new technology. Each group represents a unique *psychographic* profile—a combination of psychology and demographics that makes its marketing responses different from those of the other groups. Understanding each profile and its relationship to its neighbors is a critical component of high-tech marketing lore.

*Innovators* pursue new technology products aggressively. They sometimes seek them out even before a formal marketing program has been launched. This is because technology is a central interest in their life, regardless of what function it is performing. At root they are intrigued with any fundamental advance and often make a technology purchase simply for the pleasure of exploring the new device's properties. There are not very many innovators in any given market segment, but winning them over at the outset of a marketing campaign is key nonetheless, because their endorsement reassures the other players in the marketplace that the product does in fact work.

*Early adopters*, like innovators, buy into new product concepts very early in their life cycle, but unlike innovators, they are not technologists. Rather they are people who find it easy to imagine, understand, and appreciate the benefits of a new technology, and to relate these potential benefits to their other concerns. Whenever they find a strong match, early adopters are willing to base their buying decisions upon it. Because early adopters do not rely on well-established references in making these buying decisions, preferring instead to rely on their own intuition and vision, they are key to opening up any high-tech market segment.

The *early majority* share some of the early adopter's ability to relate to technology, but ultimately they are driven by a strong sense of practicality. They know that many of these newfangled inventions end up as passing fads, so they are content to wait and see how other people are making out before they buy in themselves. They want to see well-established references before investing substantially. Because there are so many people in this segment—roughly one-third of the whole adoption life cycle—winning their business is key to any substantial profits and growth.

The *late majority* shares all the concerns of the early majority, plus one major additional one:  Whereas people in the early majority are comfortable with their ability to handle a technology product, should they finally decide to purchase it, members of the late majority are not. As a result, they wait until something has become an established standard, and even then they want to see lots of support and tend to buy, therefore, from large, well-established companies. Like the early majority, this group comprises about one-third of the total buying population in any given segment. Courting its favor is highly profitable indeed, for while profit margins decrease as the products mature, so do the selling costs, and virtually all the R&D costs have been amortized.

Finally there are the *laggards*. These people simply don't want anything to do with new technology, for any of a variety of reasons, some personal and some economic. The only time they ever buy a technological product is when it is buried so deep inside another product—the way, say, that a microprocessor is designed into the braking system of a new car—that they don't even know it is there. Laggards are generally regarded as not worth pursuing on any other basis.

To recap the logic of the Technology Adoption Life Cycle, its underlying thesis is that technology is absorbed into any given community in stages corresponding to the psychological and social profiles of various segments within that community. This process can be thought of as a continuum with definable stages, each associated with a definable group, and each group making up a predictable portion of the whole.

## The High-Tech Marketing Model

This profile is, in turn, the very foundation of the High-Tech Marketing Model. That model says that the way to develop a high-tech market is to work the curve left to right, focusing first on the innovators, growing that market, then moving on to the early adopters, growing that market, and so on, to the early majority, late majority, and even to the laggards. In this effort, companies must use each "captured" group as a reference base for going on to market to the next group. Thus, the endorsement of innovators becomes an important tool for developing a credible pitch to the early adopters, that of the early adopters to the early majority, and so on.

The idea is to keep this process moving smoothly, proceeding something like passing the baton in a relay race or imitating Tarzan swinging from vine to well-placed vine. It is important to maintain momentum in order to create a bandwagon effect that makes it natural for the next group to want to buy in. Too much of a delay, and the effect would be something like hanging from a motionless vine—nowhere to go but down. (Actually, going down is the graceful alternative. What happens more often is a desperate attempt to re-create momentum, typically through some highly visible form of promotion, which ends up making the company look like a Tarzan frantically jerking back and forth, trying to get a vine moving with no leverage. This typically leads the other animals in the jungle just to sit and wait for him to fall.)

There is an additional motive for maintaining momentum: to

keep ahead of the next emerging technology. Portable electric typewriters were displaced by portable PCs, which in turn may someday be displaced by pen-based PCs. You need to take advantage of your day in the sun before the next day renders you obsolete. From this notion comes the idea of a *window of opportunity*. If momentum is lost, then we can be overtaken by a competitor, thereby losing the advantages exclusive to a technology leadership position—specifically, the profit-margin advantage during the middle to late stages, which is the primary source from which high-tech fortunes are made.

This, in essence, is the High-Tech Marketing Model—a vision of a smooth unfolding through all the stages of the Technology Adoption Life Cycle. What is dazzling about this concept, particularly to those who own equity in a high-tech venture, is its promise of virtual monopoly over a major new market development. If you can get there first, "catch the curve," and ride it up through the early majority segment, thereby establishing the de facto standard, you can get rich very quickly and "own" a highly profitable market for a very long time to come.

## Testimonials

Lotus 1-2-3 is a prime example of optimizing the High-Tech Marketing Model. No one has ever argued that it was the best spreadsheet program ever written. Certainly it wasn't the first, and many of the features people appreciate about it most were in fact derived directly from Visicalc, its predecessor that ran on the Apple II. But Lotus 1-2-3 was the first spreadsheet for the IBM PC, and its designers were careful to tune its performance specifically for that platform. As a result, the innovators liked Lotus 1-2-3 because it was slick and fast. Then the early adopters liked it because it allowed them to do something they had never been able to do before—what later became popularized as "what if" analysis. The early majority liked the spreadsheet because it fell into line with some very common business operations, like budgeting, sales forecasting, and project tracking. As more and more people began to use it, it became harder and harder to use anything else, including paper and pencil, so the late majority gradually fell into line. This was the tool people knew how to use. If you wanted to share a spreadsheet, it had to be in Lotus format. Thus it became so entrenched that today well over half the IBM PCs and PC compatibles with spreadsheets have Lotus 1-2-3—despite the fact that there are now numerous competitors, many of which are, feature for feature, superior products.

Astounding as this accomplishment is, many other companies have achieved a comparable status. This is what Oracle has achieved in the area of relational databases, Microsoft in PC operating systems, Hewlett-Packard in PC laser printers, and IBM in mainframe computers. It is the position that Wang once owned in word processing systems, and Cray still does in scientific supercomputing, and Tandem owns in fault-tolerant on-line transaction processing. It is what puts the bounce in the stock price of companies like Autodesk (PC CAD), Mentor Graphics (CAD for workstations and minicomputers), Intel (microprocessors), and, until recently, Ashton-Tate (PC databases).

Each of these companies has had market share in excess of 50 percent in its prime market. All of them have been able to establish strongholds in the early majority segment, if not beyond, and to look forward from that position to continued growth, wondrously strong profit margins, and increasingly preferred relationships as suppliers to their customers. To be sure, some like Wang and Ashton-Tate have fallen on hard times, but only after persistent mishaps and mismanagement. For as the recent trials of Lotus, IBM, and Oracle have shown, customers are willing to tolerate an extraordinary number of missteps from a front-runner, bringing cries of anguish from their competitors who would never be granted such grace.

It should come as no surprise that the history of these flagship products conforms to the High-Tech Marketing Model. In truth, the model was essentially derived from an abstraction of these histories. And so high-tech marketing, as we enter the 1990s, keeps before it the example of these companies and the abstraction of the High-Tech Marketing Model, and marches confidently forward.

Of course, if that were a sufficient formula for success, you would need to read no further.

## Illusion and Disillusion: Cracks in the Bell Curve

It is now time to advise you that there are any number of us in Silicon Valley who are willing to testify that there is something wrong with the High-Tech Marketing Model. We believe this to be true because we all own what once were meaningful equity stakes in corporations that either no longer exist or whose current valuation is so diluted that our stock—were there a market for it, which there is not—has lost all monetary significance.

Although we all experienced our fates uniquely, much of our shared experience can be summarized by recasting the Technology Adoption Life Cycle in the following way:

### The *Revised* Technology Adoption Life Cycle

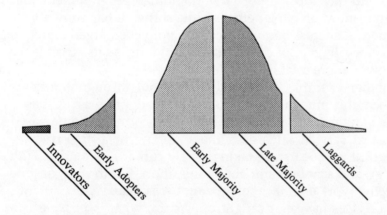

As you can see, the components of the life cycle are unchanged, but between any two psychographic groups has been introduced a gap. This symbolizes the dissociation between the two groups—that is, the difficulty any group will have in accepting a new product if it is presented in the same way as it was to the group to its immediate left. Each of these gaps represents an opportunity for marketing to lose momentum, to miss the transition to the next segment, thereby never to gain the promised land of profit-margin leadership in the middle of the bell curve.

## The First Crack

Two of the gaps in the High-Tech Marketing Model are relatively minor—what one might call "cracks in the bell curve"—yet even here unwary ventures have slipped and fallen. The first is between the innovators and the early adopters. It is a gap that occurs when a hot technology product cannot be readily translated into a major new benefit—something like Esperanto. The enthusiast loves it for its architecture, but nobody else can even figure out how to start using it.

At present, neural networking software falls into this category. Available since the 1980s, this software mimics the structure of the brain, actually programming itself through the use of feedback and rules to improve its performance against a given task. It is a terribly exciting technology because, for the first time, it holds out the possibility that computers can teach themselves and develop solutions that no human programmer could design from scratch. Nonetheless, the software has shown little commercial success because there has not yet emerged a unique and compelling application that would drive its acceptance over other, more established alternatives.

Another example of a product that fell through the crack between the innovators and the early adopters is PC-based home banking, or more generally, any form of home-based video text service, such as home shopping or airline reservations. IBM and Sears are taking a new run at this market with Prodigy, and school is still out on this effort. But to date, despite perfectly workable technology and reasonable price points, such services have not caught on. Again, what is missing is a unique and compelling application wherein video text can demonstrate itself as a form of communication substantially superior to any other alternative.

In the case of Prodigy, this much is clear: the compelling application is not home shopping, and it is probably not getting up-to-date news, even financial news. What it may be, interestingly enough, is engaging in electronic conversation. Initial market research is showing clusters of users emerging who spend a large amount of time on the bulletin boards, just talking to one another over their computers. This form of communication supports a unique combination of anonymity and intimacy that may amount to something compelling. To my way of thinking, it is not a very good bet, but it is possible.

The market-development problem in the case of both neural networking software and home-based PC networking applications is this: With each of these exciting, functional technologies it has been possible to establish a working system and to get innovators to adopt it. But it has not as yet been possible to carry that success over to the early adopters. As we shall see in the next chapter, the key to winning over this segment is to show that the new technology enables some strategic leap forward, something never before possible, which has an intrinsic value and appeal to the nontechnologist. This benefit is typically symbolized by a single, compelling application, the one thing that best captures the power and value of the new product. If the marketing effort is unable to find that compelling application, then market development stalls with the innovators, and the future of the product falls through the crack.

### The Other Crack

There is another crack in the bell curve, of approximately equal magnitude, that falls between the early majority and the late majority. By this point in the Technology Adoption Life Cycle, the market is already well developed, and the technology product has been absorbed into the mainstream. The key issue now, transitioning from the early to the late majority, has to do with demands on the end user to be technologically competent.

Simply put, the early majority is willing and able to become technologically competent, where necessary; the late majority, much less so. When a product reaches this point in the market development, it must be made increasingly easier to adopt in order to continue being successful. If this does not occur, the transition to the late majority may well stall or never happen.

The IBM PC running the Microsoft DOS operating system is currently in this situation, especially regarding penetration of the home market, the home office, and small business—none of which has the kind of in-house support resources that Fortune 2000 companies get from their MIS groups. That is one of the reasons there has been so much interest in Microsoft's Windows 3.0, an operating environment designed to make the IBM PC more Macintosh-like, hence easier to use. It is why IBM's new home product incorporates its own four-window interface that boots up automatically, so end users can use two key applications—Microsoft Works and Prodigy—without having to go through DOS.

Now millions of people have already learned how to use DOS—so why switch? The reason is that an increasing share of new customers must come from the late majority, few of whom will commit to this level of learning. Because the IBM PC and PC compatibles have been late in bringing these types of improvements to market, they have experienced a significant slowdown in U.S. sales, despite the fact that large, unsaturated market segments exist in small businesses, home offices, and home computing.

Other examples of products in danger of falling through the crack between the early and the late majority are software for desktop publishing and software for project management. Products like Pagemaker from Aldus, which pioneered the desktop publishing market, and Timeline from Symantec, which is the leader in PC-based project management packages, are simply too difficult for the late majority to adopt. And so these categories too are stagnating, although the market is nowhere near saturated.

## Discovering the Chasm

The real news, however, is not the two cracks in the bell curve, the one between the innovators and the early adopters, the other between the early and late majority. No, the real news is the deep and dividing *chasm* that separates the early adopters from the early majority. This is by far the most formidable and unforgiving transition in the Technology Adoption Life Cycle, and it is all the more dangerous because it typically goes unrecognized.

The reason the transition can go unnoticed is that with both groups the customer list and the size of the order can look the same. Typically, in either segment, you would see a list of Fortune 500 to Fortune 2000 customers making relatively large orders—five figures for sure, more often six figures or even higher. But in fact the basis for the sale—what has been promised, implicitly or explicitly, and what must be delivered—is radically different.

What the early adopter is buying, as we shall see in greater detail in Chapter 2, is some kind of *change agent*. By being the first to implement this change in their industry, the early adopters expect to get a jump on the competition, whether from lower product costs, faster time to market, more complete customer service, or some other comparable business advantage. They expect a radical discontinuity between the old ways and the new, and they are prepared to champion this cause against entrenched resistance. Being the first, they also are prepared to bear with the inevitable bugs and glitches that accompany any innovation just coming to market.

By contrast, the early majority want to buy a *productivity improvement* for existing operations. They are looking to minimize the discontinuity with the old ways. The want evolution, not revolution. They want technology to enhance, not overthrow, the established ways of doing business. And above all, they do not want to debug somebody else's product. By the time they adopt it, they want it to work properly and to integrate appropriately with their existing technology base.

This contrast just scratches the surface relative to the differences and incompatibilities among early adopters and the early majority. Let me just make two key points for now: Because of these incompatibilities, early adopters do not make good references for the early majority. And because of the early majority's concern not to disrupt their organizations, good references are critical to their buying decisions. So what we have here is a catch-22. The only suitable reference for an early majority customer, it turns out, is another member of the early majority, but no upstanding member of the early majority will buy without first having consulted with several suitable references.

## Bodies in the Chasm

What happens in this catch-22 situation? First, because the product *has* caught on with the early adopters, it has gotten a lot of publicity. Solar power—indeed, any alternative fuel source— is a good example, as are the following from the electronics industry: CASE (computer-aided software engineering), document image processing, multimedia software, IBM's OS/2 operating system, desktop devices for making 35-mm slide presentations, optical character recognition software, Steve Jobs's NeXT computer, specialized computers used solely as database servers, and video conferencing. We have all read a lot about these types of products, yet not one has achieved to date a mainstream market leadership position, despite the fact that the products actually do work reasonably well. In large part this is because of the high degree of discontinuity implicit in their adoption by organizations, and the inability of the marketing effort, to date, to lower this barrier to the early majority. So the products languish, continuing to feed off the early adopter segment of the market, but unable to really take off and break through to the high-volume opportunities.

The classic example of this scenario—finally broken through in the last couple of years—was in the local area network (LAN) marketplace with its ever-renewing "Year of the LAN." No one is quite sure when the first year of the LAN was (possibly 1984—it having, if nothing else, good Orwellian credentials). But in that first year, and for every year thereafter, its promoters solemnly pronounced that *this year* really would be the year of the LAN. After four or five unsuccessful tries, even the prognosticators themselves began giggling, but the stockholders of Ungermann-Bass, AT&T, Xerox, and 3-Com certainly weren't. The point was that LANs were in the chasm, and it wasn't until a relatively continuous version of the innovation came along—Novell software running on IBM PC hardware with interface cards that fit into existing IBM PC slots—that the early majority was finally able to buy in.

Let's look at another example. One of the great cover stories of the early 1980s was artificial intelligence (AI)—brains in a box. Everybody was writing about it, and many prestigious customer organizations were jumping on the bandwagon of companies like Teknowledge, Symbolics, and Intellicorp. Indeed, the customer list of any one of these companies looked like a Who's Who of the Fortune 100. Early AI pioneers, like Tom Kehler, the chairman of Intellicorp, routinely got coverage everywhere from *Inc.* and *High Technology* to *Time* magazine to the front page of the *Wall Street Journal*, and among other things, were able to ride that wave of enthusiasm to take their companies public.

Today, however, AI has been relegated to the trash heap. Despite the fact that it was—and is—a very hot technology, and that it garnered strong support from the early adopters, who saw its potential for using computers to aid human decision making, it has simply never caught on as a product for the mainstream market. Why? When it came time for the early majority to absorb it into the mainstream, there were too many obstacles to its adoption: lack of support for mainstream hardware, inability to integrate it easily into existing systems, no established design methodology, and a lack of people trained in how to implement it. So AI languished at the entrance to the mainstream, for lack of a sustained marketing effort to lower the barriers to adoption, and after a while it got a reputation as a failed attempt. And as soon as that happened, the term itself became taboo.

So today, although the technology of AI is alive and kicking, underlying such currently popular manifestations as so-called expert systems and object-oriented programming, no one uses the phrase *artificial intelligence* in their marketing efforts. And a company like Intellicorp, which had struggled desperately to be profitable as an AI firm, has backed completely away from that identity.

In sum, when promoters of high-tech products try to make the transition from a market base made up of visionary early adopters to penetrate the next adoption segment, the pragmatist early majority, they are effectively operating *without a reference base and without a support base within a market that is highly reference oriented and highly support oriented.*

This is indeed a chasm, and into this chasm many an unwary start-up venture has fallen. Despite repeated instances of the chasm effect, however, high-tech marketing has yet to get this problem properly in focus. Indeed, that is the function of this book. As a final prelude to that effort, therefore, by way of evoking additional glimmers of recognition and understanding of this plight of the chasm, I offer the following parable as a kind of condensation of the entrepreneurial experience gone awry.

## A High-Tech Parable

In the first year of selling a product—most of it alpha and beta release—the emerging high-tech company expands its customer list to include some technology enthusiast innovators and one or two visionary early adopters. Everyone is pleased, and at the first annual Christmas party, held on the company premises, plastic glasses and potluck canapés are held high.

In the second year—the first year of true product—the company wins over several more visionary early adopters, including a handful of truly major deals. Revenue meets plan, and everyone is convinced it is time to ramp up—especially the venture capitalists who note that next year's plan calls for a 300 percent increase in revenue. (What could justify such a number? The technology adoption profile, of course! For are we not just at that point in the profile where the slope is increasing at its fastest point? We don't want to lose market share at this critical juncture to some competitor. We must act while we are still within our window of opportunity. Strike while the iron is hot!) This year the company Christmas party is held at a fine hotel, the glasses are crystal, the wine vintage, and the theme, à la Dickens, is "Great Expectations."

At the beginning of the third year, a major sales force expansion is undertaken, impressive sales collateral and advertising are underwritten, district offices are opened, and customer support is strengthened. Halfway through the year, however, sales revenues are disappointing. A few more companies have come on board, but only after a prolonged sales struggle and significant compromise on price. The number of sales overall is far fewer than expected, and growth in expenses is vastly outdistancing growth in income. In the meantime, R&D is badly bogged down with several special projects committed to in the early contracts with the original customers.

Meetings are held (for the young organization is nothing if not participative in its management style). The salespeople complain that there are great holes in the product line and that what is available today is overpriced, full of bugs, and not what the customer wants. The engineers claim they have met spec and schedule for every major release, at which point the customer support staff merely groan. Executive managers lament that the sales force doesn't call high enough in the prospect organization, lacks the ability to communicate the vision, and simply isn't aggressive enough. Nothing is resolved, and, off line, political enclaves begin to form.

Third quarter revenue results are in—and they are absolutely dismal. It is time to whip the slaves. The board and the venture capitalists start in on the founders and the president, who in turn put the screws to the vice president of sales, who passes it on to the troops in the trenches. Turnover follows. The vice-president of marketing is fired. It's time to bring in "real management." More financing is required, with horrendous dilution for the initial cadre of investors—especially the founders and the key technical staff. One or more founders object but are shunted aside. Six months pass. Real management doesn't do any better. Key defections occur. Time to bring in consultants. More turnover. What we really need now, investors decide, is a turnaround artist. Layoffs followed by more turnover. And so it goes. When the screen fades to the credits, yet another venture rides off to join the twilight companies of Silicon Valley—enterprises on life support, not truly alive and yet, due in part to the vagaries of venture capital accounting, unable to choose death with dignity.

Now, it is possible that this parable overstates the case—I have been accused of such things in the past. But there is no overstating the case that year in and year out hundreds of high-tech start-ups, despite having good technology and exciting products, and despite initial promising returns from the market, falter and then fail. Here's why.

What the company staff interpreted as a ramp in sales leading smoothly "up the curve" was in fact an initial blip—what we will be calling the *early market*—and not the first indications of an emerging *mainstream market*. The company failed because its managers were unable to recognize that there is something fundamentally different between a sale to an early adopter and a sale to the early majority, even when the company name on the check reads the same. Thus, at a time of greatest peril, when the company was just entering the chasm, its leaders held high expectations rather than modest ones, and spent heavily in expansion projects rather than husbanding resources.

All this is the result of high-tech marketing illusion—the belief induced by the High-Tech Marketing Model that new markets unfold in a continuous and smooth way. In order to avoid the perils of the chasm, we need to achieve a new state—high-tech marketing enlightenment—by going deeper into the dynamics of the Technology Adoption Life Cycle to correct the flaws in the model and provide a secure basis for marketing strategy development.

# 2

# High-Tech Marketing Enlightenment

---

*First there is a mountain,*
*Then there is no mountain,*
*Then there is.*
　　　　　　　—Zen proverb

What is it about California? How can any state be so successful economically and yet so weird? I myself am from Oregon, a perfectly normal state, with the kind of sagging, suffering economy that states are supposed to have. A few fish, a few trees, a few jobs—that's all. I never intended to move south and write a book that says, in the very next paragraph mind you, that you should bet your next million on a Zen proverb. California is a bad influence.

However, if you are going to risk time and money in high tech, then you really do need to remember how high-tech markets develop, and the following proverb is as good a way as any:

*First there is a market* . . . Made up of innovators and early adopters, it is an early market, flush with enthusiasm and vision and, often as not, funded by a potful of dollars earmarked for accomplishing some grand strategic goal.

*Then there is no market* . . . This is the chasm period, during which the early market is still trying to digest its ambitious projects, and the mainstream market waits to see if anything good will come of them.

*Then there is.* If all goes well, and the product and your company pass through the chasm period intact, then a mainstream market does emerge, made up of the early and the late majority. With them comes the real opportunity for wealth and growth.

To reap the rewards of the mainstream market, your marketing strategy must successfully respond to all three of these stages. In each case, the key to success is to focus in on the dominant "adoption type" in the current phase of the market, learn to appreciate that type of person's psychographics, and then adjust your marketing strategy and tactics accordingly. Illustrating how to do that is the goal of this chapter.

## First Principles

Before we get started, however, we need to establish some ground rules. The first step toward enlightenment is to get a firm grasp on the obvious. In our case, that means getting a useful working definition of the word *marketing*. *Useful* in this context means actionable—can we find in the concept of marketing a reasonable basis for taking actions that will predictably and positively affect company revenues? That, after all, is the purpose of this book.

Actually, in this context, defining marketing is not particularly difficult: it simply means taking actions to create, grow, maintain, or defend markets. What a *market* is we will get to in a moment, but it is, first, a real thing, independent of any one individual's actions. Marketing's purpose, therefore, is to develop and shape something that is real, and not, as people sometimes want to believe, to create illusions. In other words, we are dealing with a discipline more akin to gardening or sculpting than, say, to spray painting or hypnotism.

Of course, talking this way about marketing merely throws the burden of definition onto *market*, which we will define, for the purposes of high tech, as

- a set of actual or potential customers
- for a given set of products or services
- who have a common set of needs or wants, and
- who reference each other when making a buying decision.

People intuitively understand every part of this definition except the last. Unfortunately, getting the last part—the notion that part of what defines a high-tech market is the tendency of its members to reference each other when making buying decisions—is absolutely key to successful high-tech marketing. So let's make this as clear as possible.

If two people buy the same product for the same reason but have no way they could reference each other, they are not part of the same market. That is, if I sell an oscilloscope for monitoring heartbeats to a doctor in Boston and the identical product for the same purpose to a doctor in Zaire, and these two doctors have no reasonable basis for communicating with each other, then I am dealing in two different markets. Similarly, if I sell an oscilloscope to a doctor in Boston and then go next door and sell the same product to an engineer working on a sonar device, I am also dealing in two different markets. In both cases, the reason we have separate markets is because the customers could not have referenced each other.

Depending on what day of the week it is, this idea seems to be either blindingly obvious or doubtful at best. Staying with the example at hand, can't one argue that there is, after all, such a thing as the oscilloscope market? Well, yes and no. If you want to use the word *market* in this sense, it stands for the aggregate sales, both past and projected, for oscilloscopes. If that is how you want to use the word—say, if you are a financial analyst—that's fine, but you had better realize you are adding apples and oranges (that is, doctor sales + engineer sales) to get your final totals, and in so doing, you are leaving yourself open to misinterpreting the data badly. Most importantly, *market*, when it is defined in this sense, ceases to be a single, isolable object of action—it no longer refers to any single entity that can be acted on—and cannot, therefore, be the focus of *marketing*.

The way around this problem for many marketing professionals is to break up "the market" into isolable "market segments." *Market segments*, in this vocabulary, meet our definition of markets, including the self-referencing aspect. When marketing consultants sell market segmentation studies, all they are actually doing is breaking out the natural market boundaries within an aggregate of current and potential sales.

Marketing professionals insist on market segmentation because they know no meaningful marketing program can be implemented across a set of customers who do not reference each other. The reason for this is simply leverage. No company can afford to pay for every marketing contact made. Every program must rely on some ongoing chain-reaction effects—what is usually called word of mouth. The more self-referencing the market and the more tightly bounded its communications channels, the greater the opportunity for such effects.

So much for first principles. There are additional elements to our final definition of market—principally, a concept called "the whole product"—but we will get to that later in the book. For now, let's apply what we have to the three phases of high-tech marketing. The first of these is the *early market*.

## Early Markets

The initial customer set for a new technology product is made up primarily of innovators and early adopters. In the high-tech industry, the innovators are better known as *technology enthusiasts* or just *techies*, whereas the early adopters are the *visionaries*. It is the latter group, the visionaries, who dominate the buying decisions in this market, but it is the technology enthusiasts who are first to realize the potential in the new product. High-tech marketing, therefore, begins with the techies.

### Innovators: The Technology Enthusiasts

Classically, the first people to adopt any new technology are those who appreciate the technology for its own sake. For readers old enough to have been raised on Donald Duck comic books from Walt Disney, Gyro Gearloose may well have been your first encounter with a technology enthusiast. Or, if you were more classically educated, perhaps it was Archimedes crying, "Eureka!" at discovering the concept of measuring specific gravity through the displacement of water, or Daedalus, inventing a labyrinth and then the wings whereby one could fly out of it (if one did not fly too close to the sun). Or, for those who turn more towards movies and TV, more familiar examples of the type include *Back to the Future's* Doc Brown or the Professor from "Gilligan's Island." "Inventors," "propeller heads," "nerds," "techies"—we have many labels for a group of people who are, as a rule and despite a tendency toward introversion, delightful companions—provided you like to talk about technical topics.

They are the ones who first appreciate the architecture of your product and why it therefore has a competitive advantage over the current crop of products established in the marketplace. They are the ones who will spend hours trying to get products to work that, in all conscience, never should have been shipped in the first place. They will forgive ghastly documentation, horrendously slow performance, ludicrous omissions in functionality, and bizarrely obtuse methods of invoking some needed function—all in the name of moving technology forward. They make great critics because they truly care.

To give some high-tech examples, technology enthusiasts are the ones who bought VCRs, compact disk players, and camcorders when they each cost well over a thousand dollars. They bought personal computers in the 1970s when you had to build them from kits. They wrote their own software and then shared it with each other over electronic bulletin board services. Today they are interested in voice synthesis and voice recognition, interactive multimedia systems, neural networks, the modeling of chaos in Mandelbrot sets, and the notion of an artificial life based on silicon. They are the primary customers for bulletin board services—especially those that discount during the late night hours—and PC-based video games, but they will be found anywhere that anyone is "pushing the edge of the envelope."

Sometimes a technology enthusiast becomes famous—usually as the inventor of a lucrative product. In the world of PCs, Bill Gates started business life this way, but he may have forfeited his status somewhat as he became more Machiavellian. Steve Wozniak (creator of the Apple II), on the other hand, has stayed in role, as have Andy Herzfeld (lead designer for Macintosh's MultiFinder), Dan Bricklin (author of Visicalc), Marc Canter (author of Macromind's Director), and Dave Winer (author of More).

My personal favorite, though, is a fellow named David Lichtman with whom I worked at Rand Information Systems in the late seventies and early eighties. Long before anyone was taking PCs seriously, David showed me one he had put together himself—including, as a peripheral, a voice synthesizer. This was sitting on his desk at work right next to a little microprocessor-driven box he had invented to fill out his time sheet for him. If you followed David home, you would find a house littered with cameras, sound equipment, and assorted electronic toys. And at work, whenever there was any question about how a particularly arcane or intricate tool actually functioned, David was the man to ask. He was the archetypal technology enthusiast.

In business, technology enthusiasts are the gatekeepers for any new technology. They are the ones who have the interest to learn about it and the ones everyone else deems competent to do the early evaluation. As such, they are the first key to any high-tech marketing effort.

As a buying population, or as key influencers in corporate buying decisions, technology enthusiasts pose fewer requirements than any other group in the adoption profile—but you must not ignore the issues that are important to them. First, and most crucially, they want the truth, and without any tricks. Second, wherever possible, whenever they have a technical problem, they want access to the most technically knowledgeable person to answer it. Often this may not be sound from a management point of view, and you will have to deny or restrict such access, but you should never forget that it is wanted.

Third, they want to be first to get the new stuff. By working with them under nondisclosure—a commitment to which they typically adhere scrupulously—you can get great feedback early in the design cycle and begin building a supporter who will influence buyers not only in his own company but elsewhere in the marketplace as well. Finally, they want everything cheap. This is sometimes a matter of budgets, but it is more fundamentally a problem of perception—they think all technology should be free or available at cost, and they have no use for "added-value" arguments. The key consequence here is, if it is their money, you have to make it available cheap, and if it is not, you have to make sure price is not their concern.

In large companies, technology enthusiasts can most often be found in the advanced technology group, or some such congregation, chartered with keeping the company abreast of the latest developments in high tech. There they are empowered to buy one of almost anything, simply to explore its properties and examine its usefulness to the corporation. In smaller companies, which do not have such budgetary luxuries, the technology enthusiast may well be the "designated techie" in the MIS group or a member of a product design team who either will design your product in or supply it to the rest of the team as a technology aid or tool.

To reach technology enthusiasts, you need to place your message in one of their various haunts—computer bulletin boards, retail storefronts that cater specifically to the technology expert, technical publications, technology conferences, and the like. Direct response advertising works well with this group, as they are the segment most likely to send for literature, or a free demo, or whatever you offer. Finally, don't waste your time with a lot of fancy image advertising—they read all that as just marketing hype. An ad in the back of the magazine will reach them—they read cover to cover.

In sum, technology enthusiasts are easy to do business with, provided you (1) have the latest and greatest technology, and (2) don't need to make much money. For any innovation, there will always be a small class of these enthusiasts who will want to try it out *just to see if it works*. For the most part, these people are not powerful enough to dictate the buying decisions of others, nor do they represent a significant market in themselves. What they represent instead is a beachhead, a source of initial product or service references, and a test bed for introducing modifications to the product or service until it is thoroughly "debugged."

In *In Search of Excellence*, for example, Peters and Waterman tell the story of the fellow who invented Post-It notes. He just put them on the desk of secretaries, and some of those secretaries just tried them to see if or how they would work. Those secretaries became Post-It note enthusiasts and were an early key in the campaign to keep the product idea alive. Enthusiasts are like kindling: they help start the fire. They need to be cherished for that. The way to cherish them is to let them in on the secret, to let them play with the product and give you the feedback, and wherever appropriate, to implement the improvements they suggest and to let them know that you implemented them.

The other key to working with enthusiasts toward a successful marketing campaign is to find the ones who are near or have access to *the big boss*. Big bosses are people who can dictate purchases and who do represent a significant marketing opportunity in and of themselves. To get more specific about the kind of big boss we are looking for, let us now turn to the next group in the Technology Adoption Life Cycle, the *early adopters*, or as they are often called in the high-tech industry, the *visionaries*.

## Early Adopters: The Visionaries

Visionaries are that rare breed of people who have the insight to match an emerging technology to a strategic opportunity, the temperament to translate that insight into a high-visibility, high-risk project, and the charisma to get the rest of their organization to buy into that project. They are the early adopters of high-tech products. Often working with budgets in the multiple millions of dollars, they represent a hidden source of venture capital that funds high-technology business.

When John F. Kennedy launched the U.S. space program, he showed himself to be something we in America had not known for some time—a visionary president. When Henry T. Ford implemented factory-line mass production of automobiles so that every family in America could afford one, he became one of our best-known business visionaries. When Steve Jobs took the Xerox PARC interface out of the laboratory and put it into a personal computer "for the rest of us," then drove the rest of the PC industry to accept this computer almost in spite of itself, he showed himself to be a visionary to be reckoned with.

As a class, visionaries tend to be recent entrants to the executive ranks, highly motivated, and driven by a "dream." The core of the dream is a business goal, not a technology goal, and it involves taking a quantum leap forward in how business is conducted in their industry or by their customers. It also involves a high degree of personal recognition and reward. Understand their dream, and you will understand how to market to them.

To give additional examples specific to high tech, when Max Hopper committed American Airlines to the Sabre System of on-line reservations, he was acting as a visionary. When Tim Turnpaugh converted SeaFirst Bank to IBM mainframes and then, to access those mainframes, installed Macintoshes instead of 3270 terminals on every desk, he was acting as a visionary. When Ed Mahler committed millions of DuPont's dollars to artificial intelligence system development, he was acting as a visionary. When, just recently, Sheldon Laube at Price Waterhouse committed to purchase and install ten thousand copies of Lotus's new and totally unproven product, Notes, he was acting as a visionary. In every case, these people took significant business risks with what at the time was unproven technology in order to achieve breakthrough improvements in productivity and customer service.

And that is the key point. Visionaries are not looking for an *improvement*; they are looking for a fundamental *breakthrough*. Technology is important only insomuch as it promises to deliver on this dream. If the dream is breakthrough reductions in inventory carrying costs through just-in-time manufacturing, then the system will be some variant of MRP II. If it is radical product-cost reductions through improved manufacturing quality and process yield, then the technology will include statistical quality control and real-time monitoring of work cell output. If it is unmatched freshness of product on the store shelf and timeliness of product sales data, it might include hand-held, portable computing devices that can upload product status to an inventory control system. If it is 24-hours-a-day customer service through automated tellers, then the technology system will have to be nonstop and fault tolerant. The key point is that, in contrast with the technology enthusiast, a visionary derives value not from a system's technology itself but from the strategic leap forward it enables.

Visionaries drive the high-tech industry because they see the potential for an "order-of-magnitude" return on investment and willingly take high risks to pursue that goal. They will work with vendors who have little or no funding, with products that start life as little more than a diagram on a whiteboard, and with technology gurus who bear a disconcerting resemblance to Rasputin. They know they are going outside the mainstream, and they accept that as part of the price you pay when trying to leapfrog the competition.

Because they see such vast potential for the technology they have in mind, they are the least price-sensitive of any segment of the technology adoption profile. They typically have budgets that let them allocate generous amounts toward the implementation of a strategic initiative. This means they can usually provide up-front money to seed additional development that supports their project—hence their importance as a source of high-tech development capital.

Finally, beyond fueling the industry with dollars, visionaries are also effective at alerting the business community to pertinent technology advances. Outgoing and ambitious as a group, they are usually more than willing to serve as highly visible references, thereby drawing the attention of the business press and additional customers to small fledgling enterprises.

As a buying group, visionaries are easy to sell but very hard to please. This is because they are buying a dream—which, to some degree, will always be a dream. The "incarnation" of this dream will require the melding of numerous technologies, many of which will be immature or even nonexistent at the beginning of the project. The odds against everything falling into place without a hitch are astronomical. Nonetheless, both the buyer and the seller can build successfully on two key principles.

First, visionaries like a project orientation. They want to start out with a pilot project, which makes sense because they are "going where no man has gone before" and you are going with them. This is followed by more project work, conducted in phases, with milestones, and the like. The visionaries' idea is to be able to stay very close to the development train to make sure it is going in the right direction and to be able to get off if they discover it is not going where they thought.

While reasonable from the customer's point of view, this project orientation is usually at odds with the entrepreneurial vendors who are trying to create a more universally applicable product around which they can build a multicustomer business. This potentially lose/lose situation—threatening both the quality of the vendor's work and the fabric of the relationship—requires careful account management, including frequent contact at the executive level.

The winning strategy is built around the entrepreneur being able to "productize" the deliverables from each phase of the visionary project. That is, whereas for the visionary the deliverables of phase one are only of marginal interest—proof of concept with some productivity improvement gained, but not "the vision"—these same deliverables, repackaged, can be a whole product to someone with less ambitious goals. For example, a company might be developing a comprehensive object-oriented software toolkit, capable of building systems that could model the entire workings of a manufacturing plant, thereby creating an order-of-magnitude improvement in scheduling and processing efficiency. The first deliverable of the toolkit might be a model of just one milling machine's operations and its environment. The visionary looks at that model as a milestone. But the vendor of that milling machine might look at the same model as a very desirable product alterations and want to license it with only modest alterations. It is important, therefore, in creating the phases of the visionary's project to build in milestones that lend themselves to this sort of product spin-off.

The other key quality of visionaries is that they are in a hurry. They see the future in terms of windows of opportunity, and they see those windows closing. As a result, they tend to exert deadline pressures—the carrot of a big payment or the stick of a penalty clause—to drive the project faster. This plays into the classic weaknesses of entrepreneurs—lust after the big score and overconfidence in their ability to execute within any given time frame.

Here again, account management and executive restraint are crucial. The goal should be to package each of the phases such that each phase

1. is accomplishable by mere mortals working in earth time

2. provides the vendor with a marketable product

3. provides the customer with a concrete return on investment that can be celebrated as a major step forward.

The last point is crucial. Getting closure with visionaries is next to impossible. Expectations derived from dreams simply cannot be met. This is not to devalue the dream, for without it there would be no directing force to drive progress of any sort. What is important is to celebrate continually the tangible and partial as both useful things in their own right and as heralds of the new order to come.

The most important principle stemming from all this is the emphasis on management of expectations. Because controlling expectations is so crucial, the only practical way to do business with visionaries is through a small, top-level direct sales force. At the front end of the sales cycle, you need such a group to understand the visionaries' goals and give them confidence that your company can step up to those. In the middle of the sales cycle, you need to be extremely flexible about commitments as you begin to adapt to the visionaries' agenda. At the end, you need to be very careful in negotiations, keeping the spark of the vision alive without committing to tasks that are unachievable within the time frame allotted. All this implies a mature and sophisticated representative working on your behalf.

In terms of prospecting for visionaries, they are not likely to have a particular job title, except that, to be truly useful, they must have achieved at least a vice presidential level in order to have the clout to fund their visions. In fact, in terms of communications, typically you don't find them, they find you. The way they find you, interestingly enough, is by maintaining relationships with technology enthusiasts. That is one of the reasons why it is so important to capture the technology enthusiast segment.

In sum, visionaries represent an opportunity early in a product's life cycle to generate a burst of revenue and gain exceptional visibility. The opportunity comes with a price tag—a highly demanding customer who will seek to influence your company's priorities directly and a high-risk project that could end in disappointment for all. But without this boost many high-tech products cannot make it to market, unable to gain the visibility they need within their window of opportunity, or unable to sustain their financial obligations while waiting for their marketplace to develop more slowly. Visionaries are the ones who give high-tech companies their first big break. It is hard to plan for them in marketing programs, but it is even harder to plan without them.

## The Dynamics of Early Markets

To get an early market started requires an entrepreneurial company with a breakthrough technology product that enables a new and compelling application, a technology enthusiast who can evaluate and appreciate the superiority of the product over current alternatives, and a well-heeled visionary who can foresee an order-of-magnitude improvement from implementing the new application. When the market is unfolding as it should, the entrepreneurial company seeds the technology enthusiast community with early copies of its product while at the same time sharing its vision with the visionary executives. It then invites the visionary executives to check with the technology enthusiast of their choice to verify that the vision is indeed achievable. Out of these conversations comes a series of negotiations in which, for what seems like a very large amount of money at the time, but which will later be recognized as just the tip of the iceberg, the technology enthusiasts get to buy more toys than they have ever dreamed of, the entrepreneurial company commits itself to product modifications and system integration services it never intended to, and the visionary has what on paper looks to be an achievable project, but which is in fact a highly improbable dream.

That's when the market unfolds as it should. That is the good scenario—good because, although it is rife with problems, they are ones that will get solved one way or another, and some level of value will be achieved all around. There are numerous other scenarios where the early market does not even get a proper start. Here are some of them:

- First problem: The company simply has no expertise in bringing a product to market. It raises insufficient capital for the effort, hires inexperienced sales and marketing people, tries to sell the product through an inappropriate channel of distribution, promotes in the wrong places and in the wrong ways, and in general fouls things up.

Remedying this kind of situation is not as hard as it may seem, provided the participants in the company are still communicating and cooperating with each other, and everyone is willing to scale back their expectations several notches.

The basis for reform is the principle that winning at marketing more often than not means being the biggest fish in the pond. If we are very small, then we must search out a very small pond indeed. To qualify as a "real pond," as we also noted before, its members must be aware of themselves as a group, that is, it must constitute a self-referencing market segment, so that when we establish a leadership position with some of its members, they will get the word out—quickly and economically—to the rest.

Of course, no single pond of a size we can dominate in the short term is large enough to provide a sustaining market for the long term. Sooner or later, we have to go "pond hopping." Or, to shift the metaphor, we need to reframe our "pond" tactics in the context of a "bowling pin" strategy, where one targets a given segment not just because one can "knock it over" but because, in so doing, it will help knock over the next target segment, and thus lead to market expansion. With the right kind of angle of attack, it is amazing how large and fast the chain reaction can be. So one is never necessarily out of the game, even when things are pretty bleak.

- A second problem: The company sells the visionary before it has the product. This is a version of the famous *vaporware* problem, based on preannouncing and premarketing a product that still has significant development hurdles to overcome. At best, the entrepreneurial company secures a few pilot projects, but as schedules continue to slip, the visionary's position in the organization weakens, and support for the project is eventually withdrawn, despite a lot of customized work, with no usable customer reference gained.

Caught in this situation, the entrepreneurial company has only one adequate response, a truly unhappy one: shut down its marketing efforts, admit its mistakes to its investors, and focus all its energies into turning its pilot projects into something useful, first in terms of a deliverable to the customer, and ultimately in terms of a marketable product. Since most entrepreneurial companies are fueled, as much as anything, on the ego of their founders, this is too often the road not taken, thereby keeping bankruptcy lawyers—and, sadder still, frequently divorce lawyers—in full employment.

- Problem number three: Marketing falls prey to the crack between the technology enthusiast and the visionary by failing to discover, or at least failing to articulate, the compelling application that provides the order-of-magnitude leap in benefits. A number of companies buy the product to test it out, but it never gets incorporated into a major system rollout, because the rewards never quite measure up to the risks. The resulting lack of revenue leads to folding the effort, either by shutting it down entirely, or selling it off "for scrap" to another enterprise.

The corrective response here begins with reevaluating what we have. If it is not, in fact, a breakthrough product, then it is not ever going to create an early market. But perhaps it could serve as a supplementary product in an existing mainstream market. If that is indeed the case, then the right response is to swallow our pride, reduce our financial expectations, and subordinate ourselves to an existing mainstream-market company, who can put our product in play through its existing channels. Computer Associates, one of the largest software companies in the world, was built up almost entirely on this principle of remarketing other companies' often cast-off products.

Alternatively, if we truly have a breakthrough product, but we are stalled in getting the early market moving, then we have to step down from the lofty theoretical plateau on which we have established that this product can be part of any number of exciting applications. Then we must get very practical about focusing on one application, making sure that it is indeed a compelling one for at least one visionary who is already familiar with us, and then committing to that visionary, in return for his or her support, to removing every obstacle to getting that application adopted.

These are some of the most common ways in which an early market development effort can go off—and be put back on—track. For the most part, the problems are solvable because there are always multiple options at the outset of anything. The biggest problem is typically overly ambitious expectations combined with undercapitalization—or, as my grandmother used to put it, when your eyes are bigger than your stomach. Things get a lot more complex when we are dealing with the dynamics of mainstream markets, to which we shall now turn.

## Mainstream Markets

Mainstream markets in high tech look a lot like mainstream markets in any other industry, particularly those that sell business to business. They are dominated by the early majority, who in high tech are best understood as *pragmatists*, who, in turn, tend to be accepted as leaders by the late majority, best thought of as *conservatives*, and rejected as leaders by the laggards, or *skeptics*. As in the previous chapter, we are going to look closely at how the psychographics of each of these groups influences the development and dynamics of a high-tech market.

### Early Majority: The Pragmatists

Throughout the 1980s, the early majority, or pragmatists, have represented the bulk of the market volume for any technology product. You can succeed with the visionaries, and you can thereby get a reputation for being a high flyer with a hot product, but that is not ultimately where the dollars are. Instead, those funds are in the hands of more prudent souls, who do not want to be pioneers ("Pioneers are people with arrows in their backs"), who never volunteer to be an early test site ("Let somebody else debug your product"), and who have learned the hard way that the "leading edge" of technology is all too often the "bleeding edge."

Who are the pragmatists? Actually, important as they are, they are hard to characterize because they do not have the visionary's penchant for drawing attention to themselves. They are not the Hamlets but the Horatios, characters like Sherman Potter, the colonel on "M*A*S*H*", or Captain Bogamil in *Beverly Hills Cop*. Never the standout, they are what makes for the continuity, so that after the star either dies (tragedy) or rides off into the sunset (heroic romance, comedy), they are left to clean up and to answer the inevitable final question: Who was that masked man?

In the realm of high tech, pragmatist CEOs are not common, and those there are, true to their type, keep a low profile. Rod Canion of Compaq, Bill Campbell of GO, and Pete Petersen of WordPerfect are three that come to mind. In the role of president, John Shirley of Microsoft filled this bill admirably. The point is, pragmatists are best known by their closest colleagues, from whom they typically have earned the highest respect, and by their peers within their industry, where they show up near the top of the leader board year after year.

Of course, to market successfully to pragmatists, one does not have to be one—just understand their values and work to serve them. To look more closely into these values, if the goal of visionaries is to take a quantum leap forward, the goal of pragmatists is to make a percentage improvement—incremental, measurable, predictable progress. If they are installing a new product, they want to know how other people have fared with it. The word *risk* is a negative word in their vocabulary—it does not connote opportunity or excitement but rather the chance to waste money and time. They will undertake risks when required, but they first will put in place safety nets and manage the risks very closely.

The Fortune 2000 MIS community, as a group, is led by people who are largely pragmatist in orientation. Business demands for increased productivity push them toward the front of the adoption life cycle, but natural prudence and budget restrictions keep them cautious. As individuals, pragmatists held back from buying PCs until they saw that WordStar and Lotus really worked, held back from DEC until the benefits of

VAX compatibility and DECNet networking were incontrovertible, and are now trying to decide just how far they can go with Sun Microsystems.

If pragmatists are hard to win over, they are loyal once won, often enforcing a company standard that requires the purchase of your product, and only your product, for a given requirement. This focus on standardization is, well, pragmatic, in that it simplifies internal service demands. But the secondary effects of this standardization—increasing sales volumes and lowering the cost of sales—is dramatic. Hence the importance of pragmatists as a market segment.

The most celebrated example and beneficiary of this effect has been IBM. We tend to think of IBM's dominance as exclusive today, but actually, when the mainframe market was at its height, it supported a variety of vendors, each with its own pragmatist enclave. In the banking community, they gravitated to Burroughs; in the federal government, Honeywell; in engineering and science, Control Data; in point-of-sale systems, NCR. This was followed in the 1980s by new players who brought minicomputers to market and gained pragmatist followings, including H-P in the factories, DEC in the engineering and design groups, Tandem with the banks and on Wall Street, and Wang in the legal community. As power moved down to the desktop, IBM reentered the fray with a dominant PC for the business and financial community, while Apple became the standard for education and the graphic arts, and Sun and Apollo for a new generation of CAD systems. Each one of these companies rode a pragmatist wave within a specific market to boost its sales a quantum leap upward. It is crucial, therefore, for any long-term strategic marketing plan to understand the pragmatist buyers and to focus on winning their trust.

When pragmatists buy, they care about the company they are buying from, the quality of the product they are buying, the infrastructure of supporting products and system interfaces, and the reliability of the service they are going to get. In other words, they are planning on living with this decision personally for a long time to come. (By contrast, the visionaries are more likely to be planning on implementing the great new order and then using that as a springboard to their next great career step upward.) Because pragmatists are in it for the long haul, and because they control the bulk of the dollars in the marketplace, the rewards for building relationships of trust with them are very much worth the effort.

Pragmatists tend to be "vertically" oriented, meaning that they communicate more with others like themselves within their own industry than do technology enthusiasts and early adopters, who are more likely to communicate "horizontally" across industry boundaries in search of kindred spirits. This means it is very tough to break into a new industry selling to pragmatists. References and relationships are very important to these people, and there is a kind of catch-22 operating: pragmatists won't buy from you until you are established, yet you can't get established until they buy from you. Obviously, this works to the disadvantage of start-ups and, conversely, to the great advantage of companies with established track records. On the other hand, once a start-up has earned its spurs with the pragmatist buyers within a given vertical market, they tend to be very loyal to it, and even go out of their way to help it succeed. When this happens, the cost of sales goes way down, and the leverage on incremental R&D to support any given customer goes way up. That's one of the reasons pragmatists make such a great market.

There is no one distribution channel preferred by pragmatists, but they do want to keep the sum total of their distribution relationships to a minimum. This allows them to maximize their buying leverage and maintain a few clear points of control, should anything go wrong. In some cases this prejudice can be overcome if the pragmatist buyer knows a particular salesperson from a previous relationship. As a rule, however, the path into the pragmatist community is smoother if a smaller entrepreneurial vendor can develop an alliance with one of the already accepted vendors or if it can establish a value-added-reseller (VAR) sales base. VARs, if they truly specialize in the pragmatist's particular industry, and if they have a reputation for delivering quality work on time and within budget, represent an extremely attractive type of solution to a pragmatist. They can provide a "turnkey" answer to a problem, without impacting internal resources already overloaded with the burdens of ongoing system maintenance. What the pragmatist likes best about VARs is that they represent a single point of control, a single company to call if anything goes wrong.

One final characteristic of pragmatist buyers is that they like to see competition—in part to get costs down, in part to have the security of more than one alternative to fall back on, should anything go wrong, and in part to assure themselves they are buying from a proven market leader. This last point is crucial: pragmatists want to buy from proven market leaders because they know that third parties will design supporting products around a market-leading product. That is, market-leading products create an *aftermarket* that other vendors service. This radically reduces pragmatist customers' burden of support. By contrast, if they mistakenly choose a product that does not become the market leader, but rather one of the also-rans, then this highly valued aftermarket support does not develop, and they will be stuck making all the enhancements by themselves. Market leadership is crucial, therefore, to winning pragmatist customers.

Pragmatists are reasonably price-sensitive. They are willing to pay a modest premium for top quality or special services, but in the absence of any special differentiation, they want the best deal. That's because, having typically made a career commitment to their job and/or their company, they get measured year in and year out on what their operation has spent versus what it has returned to the corporation.

Overall, to market to pragmatists, you must be patient. You need to be conversant with the issues that dominate their particular business. You need to show up at the industry-specific conferences and trade shows they attend. You need to be mentioned in articles that run in the magazines they read. You need to be installed in other companies in their industry. You need to have developed applications for your product that are specific to the industry. You need to have partnerships and alliances with the other vendors who serve their industry. You need to have earned a reputation for quality and service. In short, you need to make yourself over into the obvious supplier of choice.

This is a long-term agenda, requiring careful pacing, recurrent investment, and a mature management team. One of its biggest payoffs, on the other hand, is that it not only delivers the pragmatist element of the Technology Adoption Life Cycle but tees up the conservative element as well. Sadly, however, high-tech industry has, for the most part, not seen fit to reap the rewards it has so carefully sown. To see how this has come about, let us now take a closer look at the conservatives.

### Late Majority: The Conservatives

The mathematics of the Technology Adoption Life Cycle model says that for every pragmatist there is a conservative. Put another way, conservatives represent approximately one-third of the total available customers within any given Technology Adoption Life Cycle. As a marketable segment, however, they are rarely developed as profitably as they could be, largely because high-tech companies are not, as rule, in sympathy with them.

Conservatives, in essence, are against discontinuous innovations. They believe far more in tradition than in progress. In the world of football coaches, for example, if one were to call Sam Wyche of the Cincinnati Bengals a technology enthusiast (for his fondness for gimmickry, as in his latest no-huddle offense to draw a 12-men-on-the-field penalty from his opponents), Bill Walsh of the San Francisco 49ers a visionary (for that is what the label "the Genius" was all about), and perhaps John Robinson of the Los Angeles Rams or Bill Parcells, formerly of the New York Giants, pragmatists, then Chuck Knox, currently of the Seattle Seahawks, Mike Ditka of the Chicago Bears, or the late George Allen of Ram and Washington Redskin acclaim would be your bedrock conservatives. And the leader of them all, for charm as well as conservatism, would be John Madden, the only football commentator in the world who travels from game to game by bus instead of by plane. All these coaches believe far more in the team's spirit than in any abstract system of X's and O's.

And that is pretty much how the conservatives in the Technology Adoption Life Cycle feel about high-tech products. The conservatives buy and use them not because of any real belief in them but because they feel they must just to stay on par with the rest of the world. But just because they use such products, they don't necessarily have to like them.

The truth is, conservatives often fear high tech a little bit. Therefore, they tend to invest only at the end of a technology life cycle, when products are extremely mature, market-share competition is driving low prices, and the products themselves can be treated as commodities. Often their real goal in buying high-tech products is simply not to get stung. Unfortunately, because they are working the low-margin end of the market, where there is little motive for the seller to build a high-integrity relationship with the buyer, they often do get stung. This only reinforces their disillusion with high tech and resets the buying cycle at an even more cynical level.

If high-tech businesses are going to be successful over the long term, they must learn to break this vicious circle and establish a reasonable basis for conservatives to want to do business with them. They must understand that conservatives do not have high aspirations about their high-tech investments and hence will not support high price margins. Nonetheless, through sheer volume, they can offer great rewards to the companies that serve them appropriately.

It is easy to understand conservatives if you can observe some aspect of their buying behavior within yourself. In my case, I am a late adopter of many kinds of consumer products. I did not buy a VCR until the price fell to under $350. I still buy my home appliances from Sears, including audio and TV equipment, and only when they break down do I think of replacing them. Although the Beatles became a favorite rock group of mine, I disliked every one of their albums the first time I heard it. I'm uncomfortable with most types of personal finance transactions, and I'm a very late adopter of any new kind of investment opportunity. Getting in touch with feelings like these helps one market to conservatives.

Conservatives like to buy preassembled packages, with everything bundled, at a heavily discounted price. The last thing they want to hear is that the software they just bought doesn't support the printer they have installed. They want high-tech products to be like refrigerators—you open the door, the light comes on automatically, your food stays cold, and you don't have to think about it. The products they understand best are those dedicated to a single function—word processors, calculators, copiers, and fax machines. The notion that a single computer could do all four of these functions does not excite them—instead, it is something they find vaguely nauseating.

The conservative marketplace provides a great opportunity, in this regard, to take low-cost, trailing-edge technology components and repackage them into single-function systems for specific business needs. The quality of the package should be quite high because there is nothing in it that has not already been thoroughly debugged. The price should be quite low because all the R&D has long since been amortized, and every bit of the manufacturing learning curve has been taken advantage of. It is, in short, not just a pure marketing ploy but a true solution for a new class of customer.

There are two keys to success here. The first is to have thoroughly thought through the "whole solution" to a particular target end user market's needs, and to have provided for every element of that solution within the package. This is critical because there is no profit margin to support an afterpurchase support system. The other key is to have lined up a low-overhead distribution channel that can get this package to the target market effectively.

Conservatives have enormous value to high-tech industry in that they greatly extend the market for high-tech components that are no longer state-of-the-art. The fact that the United States has all but conceded great hunks of this market to the Far East is testimony not so much to the cost advantages of offshore manufacturing as to the failure of onshore product planning and marketing imagination. Many Far East solutions today still bring only one value to the table—low cost. That is, they are nowhere near the goal of being a "whole product solution." Thus, they typically have to go through a VAR channel in order to be upgraded to the kind of complete system that a conservative can purchase. The difficulty in this distribution strategy is that few VARs are large enough to achieve the volume needed to leverage a conservative market. Far more dollars could be mined from this segment of the high-tech marketplace if American leading-edge manufacturers and marketers, with their high-volume channels and vast purchasing resources, simply paid more attention to it.

So, the conservative market is still something that high tech has in its future more than in its past. To be sure, a few companies have staked out their claims. Zilog, which led the microprocessor wars with the four-bit and eight-bit parts, still focuses on this technology, serving a whole host of markets for low-cost embedded microcontrollers—things that make your seat belts go beep when they aren't fastened and your VCR able to turn on at 2:00 A.M. to capture that fifties flick you have been promising yourself to see for years. Casio, Sharp, and Texas Instruments still milk markets out of calculator chips that haven't been redesigned for the better part of a decade. And the Apple II, despite five or more years of technical obsolescence, is only just now losing its position as the standard in kindergarten through sixth-grade education. Finally, the most recent new star on this horizon appears to be the Macintosh Classic, which has been so much in demand that Apple was still trying to catch up to it six months after introduction.

Despite such successes, one has the feeling that the conservative market is perceived more as a burden than an opportunity. High-tech business success within it will require a new kind of marketing imagination linked to a less venturesome financial model. The dollars are there for the making if we can meet new challenges that are as yet only partially familiar. However, as the cost of R&D radically escalates, companies are going to have to amortize that cost across bigger and bigger markets, and this must inevitably lead to the long ignored "back half" of the technology adoption curve.

## The Dynamics of Mainstream Markets

Just as the visionaries drive the development of the early market, so do the pragmatists drive the development of the mainstream market. Winning their support is not only the point of entry but the key to long-term dominance. But having done so, you cannot take the market for granted.

To maintain leadership in a mainstream market, you must at least keep pace with the competition. It is no longer necessary to be the technology leader, nor is it necessary to have the very best product. But the product must be *good enough*, and should a competitor make a major breakthrough, you have to make at least a catch-up response.

No one is better at playing this game than Oracle Corporation and its president, Larry Ellison. Oracle won the pragmatist market away from Relational Technology Inc. (now called ASK Ingres) by virtue of a single brilliant move—standardizing on SQL as its interface language. Because IBM was driving SQL to become a standard, Oracle could ride their coattails. But then it went one step further. It ported Oracle—and the SQL interface—to every piece of hardware it could find, something IBM could not, or at least certainly would not, consider doing. This appeared to solve what was rapidly becoming the single biggest headache for pragmatists—the proliferation of incompatible systems that, sooner or later, would have to be made to communicate with each other. Everyone and his mother wanted to be the glue to hold these systems together; Oracle won the job.

Having done so, however, Oracle did not sit back and take the market for granted. The independent software vendors at Ingres, Informix, and Sybase, not to mention the in-house database groups at IBM, DEC, Tandem, and Hewlett-Packard, were coming after them. Ingres came out with Ingres/Net and Ingres/Star to provide the data communications gateways to link the incompatible systems. Oracle responded with SQL*Star and SQL*Net. Did they actually have comparable products? No, but neither really did Ingres—they had preannounced. By the time real product started getting to real customers, Oracle was already on the way to closing the technology gap. Besides, most pragmatists did not want to undertake all this linking right away—they just wanted to know there was a growth path in the works. As long as Oracle could demonstrate that, they could keep the leadership position.

Along came Sybase and took a technology leadership position in something called distributed on-line transaction processing based on client/server architecture. Never mind that at present this is a relatively small niche market: it is technologically where the future of database systems is headed. Once again, Oracle did not ignore this potential threat. It announced its own client/server architecture—indeed, claimed it had had it all along—and again, just the *plan* for such a topology has been enough to keep the mainstream market reasonably well under control.

Of course, Oracle did not invent this sort of strategy. The credit there, at least for high tech, has to go to its greatest practitioner, IBM. Nor is the market so naive that Oracle can get away with empty promises anymore. Indeed, recent events have punished the company severely. Nonetheless, the company has overwhelming market share, and if it is willing to pull its horns in a bit and repair some of its fences, it should dominate the relational database business over the remaining life of the technology.

Few companies have been as competitive as Oracle. Indeed, several have shown how you can lose a mainstream market, despite everyone's efforts, including your own customers', to prevent that. Here are some of the more common ways:

- Stop investing in the market, cease funding R&D to match the competition, and milk it for money to invest elsewhere. This is what MicroPro did with WordStar, which once was the dominant PC word processing package. For three years it did not come out with a new release of its product. During that time WordPerfect was able to come out with a superior product and superior customer support and simply take the market away. You would think that the only way this can happen is for the market leader to be asleep at the switch. But in fact even that is impossible, because mainstream customers will scream loudly long before they actually switch. They don't want to switch; they just want you to respond.

  No, the only way this can happen is for company management to believe it is doing the right thing. And indeed it was during this period of the PC software industry's history when the prevailing wisdom was that you could not afford to be a one-product company. MicroPro made a conscious decision to invest in other products. It turned out that the prevailing wisdom and MicroPro's actions were wrong, something that Lotus appears to have learned just in the nick of time. But the key point is, once you are a mainstream market leader, you have to make mistakes, big mistakes, and persist in them, over the sustained objections of your best customers, in order not to lose out to a new competitor.

- Shoot yourself in the flagship. The easiest way to do this is to decide to redesign your flagship product from the ground up. Both Lotus and Ashton-Tate embarked on this course several years ago. Both achieved new records in missing announced shipping dates. Both in so doing lost credibility with their best customers. But at least when the Lotus products arrived, they worked. What greeted Ashton-Tate were headlines like, "DBase IV, Release 1.0: Version from Hell." That led to another—slow—round of rewrites. The consequences cost the CEO, Ed Esber, his job, and led virtually the whole PC industry to write off Ashton-Tate.

Even now, though, I think the ship could be saved. Mainstream customers truly abhor discontinuous innovations. Switching to another PC database will be incredibly disruptive to their operations, whether they are an in-house operation or a VAR of systems that incorporate dBase. If Ashton-Tate can steady its course even at this late date, customers will still have an interest in sticking with the firm.

MicroPro, Lotus, and Ashton-Tate all made serious marketing errors that jeopardized their mainstream-market leadership positions. Why? Part of the answer has to be that they were thinking with the wrong marketing model. Their decisions showed them to be overly focused on what was going on in early markets and too inattentive to the underdeveloped elements in the mainstream. In particular, they were not paying any attention at all to extending their market share to incorporate more conservatives.

That is changing. Lotus, for one, has a division now devoted to its Release 2.2 of the 1-2-3 spreadsheet, the release that is *not* state-of-the-art, *not* rewritten from scratch, but instead is just an upgrade of the old standard. They want to take this product into small business, where many of the PC marketplace's conservatives live. Microsoft has an entry-level systems division, built around a product called Works, an integrated word processor, spreadsheet, and database all in one, which they believe is appropriate for the PC conservative. IBM agrees, for they have packaged it into their offering for the home (read "conservative") market. (Interestingly enough, Lotus just acquired AlphaWorks, a comparable product to be remarketed as

LotusWorks, to go after the same market.)

The fact is, however, that all these efforts have come into being only recently, and the PC industry is just now feeling its way toward successful marketing techniques for the conservative customer. The key is to make a smooth transition from the pragmatist to the conservative market segments. This means maintaining a strong relationship with the pragmatists by staying competitive with alternative products, while at the same time introducing new features that further reduce the barriers to entry for a conservative.

In this regard, if we now look back over the first four profiles in the Technology Adoption Life Cycle, we see an interesting trend. The importance of the product itself, its unique functionality, when compared to the importance of the ancillary services to the customer, is at its highest with the technology enthusiast, and, at its lowest with the conservative. This is no surprise, since one's level of involvement and competence with a high-tech product are a prime indicator of when one will enter the Technology Adoption Life Cycle. The key lesson is that the longer your product is in the market, the more mature it becomes, and the more important the service element is to the customer. Conservatives, in particular, are extremely service oriented.

Now, it would be a much simpler world if conservatives were willing to pay for all this service they require. But they are not. So the corollary lesson is, we must use our experience with the pragmatist customer segment to identify all the issues that require service and then design solutions to these problems directly into the product. This must be the focus of mature market R&D—not the extension of functionality, not the massive rewrite from the ground up, but the gradual incorporation into the product of all the little aids that people develop, often on their own, to help them cope with its limitations. This is service indeed, for the best service, both from the point of view of convenience to the customer and low cost to the vendor, is no service at all.

Not following this path makes us vulnerable to the crack in the bell curve that separates pragmatists from conservatives. The latter are not anxious to admit to their pragmatist friends that they are unwilling or unable to step up to the same level of technological self-support, but that is in fact one of the key differentiating factors between the two groups. To date, high tech has not widely acknowledged this gap, with the result that the industry has experienced product life cycles that are far shorter than need be, and revenue streams that are far more dependent on the success of new products, and hence far more volatile, than in other industries.

All that being said, it is not high tech's inability to transition its marketing efforts effectively between the pragmatists and the conservatives that poses the greatest threat to its well-being. That honor, as we shall see in the next chapter, goes to another transition in the Technology Adoption Life Cycle, the place where high-tech fortunes truly are made or lost, crossing the chasm between the early market with its visionaries and the mainstream market with its pragmatists. Before passing on, however, to our main theme, there is one last element in the Technology Adoption Life Cycle that deserves at least a passing comment.

## Laggards: The Skeptics

Skeptics—the group that makes up the last one-sixth of the Technology Adoption Life Cycle—do not participate in the high-tech marketplace, except to block purchases. Thus, the primary function of high-tech marketing in relation to skeptics is to neutralize their influence. In a sense, this is a pity because skeptics can teach us a lot about what we are doing wrong— hence this postscript.

One of the favorite arguments of skeptics is that the billions of dollars invested in office automation have not improved the productivity of the office place one iota. Actually, some fairly good data exist to support this notion. Nonetheless, as you might expect, this argument outrages high-tech supporters, who can point to any number of obvious ways in which the industry eliminates or facilitates routine—or even nonroutine— office chores. But what if, instead of rushing to rebuttal, marketing were to explore the merits of the skeptic's argument?

What we might find, for example, is that while high-tech products do give time back to the individual, the individual does not necessarily give that time back to the corporation. Or we might find that the capabilities designed and manufactured into the system at great expense remain buried in the system because the user never learns about them. Or we might find that for every individual who can transform the hours invested in training into competence in a high-tech product, there might be another who cannot. The loss associated with these people is high, considering not just their time spent in training, along with any trainer time, but also the cost of the system bought to support them, the system they cannot effectively use.

The point is, as any experienced seller of high-tech products can tell you, cost-justification of high-tech purchases is a shaky venture at best. There is always the potential to return significant dollars, but it always depends on factors beyond the system itself. Put another way, this simply means that the claims that salespeople made for high-tech products are really claims made for "whole product solutions" that incorporate elements well beyond whatever high-tech manufacturers ship inside their boxes. If high-tech marketers do not take responsibility for seeing that the whole product solution is being delivered, then they are giving the skeptic an opening to block the sale. (For all the reasons just cited, the significance of whole product solutions is discussed at length later as the key component of successfully crossing the chasm and entering into the mainstream.)

What skeptics are struggling to point out is that new systems, for the most part, don't deliver on the promises that were made at the time of their purchase. This is not to say they do not end up delivering value, but rather that the value they actually deliver is not often anticipated at the time of purchase. If this is true—and to some degree I believe it is—it means that committing to a new system is a much greater act of faith than normally imagined. It means that the primary value in the act derives more from such notions as supporting a bias toward action than from any quantifiable packet of cost-justified benefits. The idea that the value of the system will be discovered rather than known at the time of installation implies, in turn, that product flexibility and adaptability, as well as ongoing account service, should be critical components of any buyer's evaluation checklist.

Ultimately the service that skeptics provide to high-tech marketers is to point continually to the discrepancies between the sales claims and the delivered product. These discrepancies, in turn, create opportunities for the customer to fail, and such failures, through word of mouth, will ultimately come back to haunt us as lost market share. Steamrolling over the skeptics, in other words, may be a great sales tactic, but it is a poor marketing one. From a marketing point of view, we are all subject to the "Emperor's New Clothes" syndrome, but particularly so in high tech, where every player in the market has a vested interest in boosting the overall perception of the industry. Skeptics don't buy our act. We ought to take advantage of that fact.

## Back to the Chasm

As the preceding pages indicate, there is clearly a lot of value in the Technology Adoption Life Cycle as a marketing model. By isolating the psychographics of customers  based on when they tend to enter the market, it gives clear guidance on how to develop a marketing program for an innovative product.

The basic flaw in the model, as we have said, is that it implies a smooth and continuous progression across segments over the life of a product, whereas experience teaches just the opposite. Indeed, making the marketing and communications transition between any two adoption segments is normally excruciatingly awkward because you must adopt new strategies just at the time you have become most comfortable with the old ones.

The biggest problem during this transition period is the lack of a customer base that can be referenced at the time of making the transition into a new segment. As we saw when we redrew the Technology Adoption Life Cycle, the spaces between segments indicate the credibility gap that arises from seeking to use the group on the left as a reference base to penetrate the segment on the right.

In some cases, the basic affinities of the market keep groups relatively close together. Early adopting visionaries, for example, tend to keep in touch with and respect the views of technology enthusiasts; this is because they need the latter to serve as a reality check on the technical feasibility of their vision and to help evaluate specific products. As a result, enthusiasts can speak to at least some of the visionaries' concerns.

In a comparable way, conservatives look to pragmatists to help lead them in their technology purchases. Both groups like to see themselves as members of a particular industry first, businesspeople second, and purchasers of technology third. Pragmatists, however, have more confidence in technology as a potential benefit and in their ability to make sound technology purchases. Conservatives are considerably more nervous about both. They are willing to go along, up to a point, with pragmatists they respect, but they are still slightly unnerved by pragmatists' overall self-confidence. So, once again, the reference base has partial value in transitioning between adoption segments.

The significance of this weakening in the reference base traces back to the fundamental point made about markets in the introduction: namely, that markets—particularly high-tech markets—are made up of people who reference each other during the buying decision. As we move from segment to segment in the technology adoption life cycle, we may have any number of references built up, *but they may not be of the right sort.*

Nowhere is this better seen than in the transition between *visionaries* and *pragmatists.* If there are to some extent minor gaps between the other adoption groups, between visionaries and pragmatists there is a great—and to a large extent, greatly ignored—chasm.

If we look deep into that chasm, we see four fundamental characteristics of visionaries that alienate pragmatists.

1. *Lack of respect for the value of colleagues' experiences.* Visionaries are the first people in their industry segment to see the potential of the new technology. Fundamentally, they see themselves as smarter than their opposite numbers in competitive companies—and, quite often, they are. Indeed, it is their ability to see things first that they want to leverage into a competitive advantage. That advantage can only come about if no one else has discovered it. They do not expect, therefore, to be buying a well-tested product with an extensive list of industry references. Indeed, if such a reference base exists, it may actually turn them off, indicating that for this technology, at any rate, they are

already too late.

Pragmatists, on the other hand, deeply value the experience of their colleagues in other companies. When they buy, they expect extensive references, and they want a good number to come from companies in their own industry segment. This, as we have already noted, creates a catch-22 situation: since there are usually only one or two visionaries per industry segment, how can you accumulate the number of references a pragmatist requires, when virtually everyone left to call on is also a pragmatist?

2. *Taking a greater interest in technology than in their industry.* Visionaries are defining the future. You meet them at technology conferences or at forums on the year 2000 and beyond. They are easy to strike up a conversation with, and they understand and appreciate what high-tech companies and high-tech products are trying to do. They want to talk ideas with bright people. They are bored with the mundane details of their own industries. They like to talk and think high tech.

   Pragmatists, on the other hand, don't put a lot of stake in futuristic things. They see themselves more in present-day terms, as the people devoted to making the wheels of their industry turn. Therefore, they tend to invest their convention time in industry-specific forums discussing industry-specific issues. Where pragmatists are concerned, sweeping changes and global advantages may make for fine speeches, but not much else.

3. *Failing to recognize the importance of existing product infrastructure.* Visionaries are building systems from the ground up. They are incarnating their vision. They do not expect to find components for these systems lying around. They do not expect standards to have been established—indeed, they are planning to set new standards. They do not expect support groups to be in place, procedures to have been established, or third parties to be available to share in the workload and the responsibility.

   Pragmatists expect all these things. When they see visionaries going their own route with little or no thought of connecting with the mainstream practices in their industry, they shudder. Pragmatists have based their careers on such connections. Once again, it is painfully obvious that visionaries, as a group, make a very poor reference base for pragmatists.

4. *Overall disruptiveness.* From a pragmatist's point of view, visionaries are the people who come in and soak up all the budget for their pet projects. If the project is a success, they take all the credit, while the pragmatists get stuck trying to maintain a system that is so "state-of-the-art" no one is quite sure how to keep it working. If the project fails, visionaries always seem to be a step ahead of the disaster, getting out of town while they can, and leaving the pragmatists to clean up the mess.

   Visionaries, successful or not, don't plan to stick around long. They see themselves on a fast track that has them leapfrogging up the corporate ladder and across corporations. Pragmatists, on the other hand, tend to be committed long term to their profession and the company at which they work. They are very cautious about grandiose schemes because they know they will have to live with the results.

All in all, it is easy to see why pragmatists are not anxious to reference visionaries in their buying decisions. Hence the chasm. This situation can be further complicated if the high-tech company, fresh from its marketing successes with visionaries, neglects to change its sales pitch. Thus, the company may be trumpeting its recent success at early test sites when what the pragmatist really wants to hear about are up-and-running production installations. Or the company may be saying "state-of-the-art" when the pragmatist wants to hear "industry standard."

The problem goes beyond pitches and positioning, though. It is fundamentally a problem of time. The high-tech vendor wants—indeed, needs—the pragmatist to buy now, and the pragmatist needs—or at least wants—to wait. Both have absolutely legitimate positions. The fact remains, however, that somewhere a clock has been started, and the question is, who is going to blink first?

For everyone's sake, it had better be the pragmatist. How to make sure of this outcome is the subject of the next section.

# PART II
## CROSSING THE CHASM

# Chapter 3

# The D-Day Analogy

The chasm is, by any measure, a very bad place to be. It promises few, if any, new customers—only those who have somehow got off the safe thoroughfares. But it does house all sorts of unpleasant folk, from disenchanted current customers to nasty competitors to unsavory investors. Their efforts conspire to tax the reserves of the fledgling enterprise seeking to pass through to the mainstream. We need to look briefly at these challenges so we can be alert in our defenses against them.

## The Perils of the Chasm

Let's begin with the lack of new customers. As opportunities from the early market of visionaries become increasingly saturated (with big-ticket products this can be after as few as 5 to 10 contracts), and with the mainstream market of pragmatists nowhere near the comfort level they need in order to buy, there is simply an insufficient marketplace of available dollars to sustain the firm. Having flirted with going cash-flow positive (especially during the months following one of the early market big orders), the trend is now reversed, and the enterprise is accelerating into increasing negative cash flow. Worse still, mainstream competitors, who up to this time had paid no attention to the fledgling entry into their market, now have caught sight of a new target, experienced one or two major losses, and set their sales forces in motion to counterattack.

There are few opportunities for refuge. Managers would like to retreat into their existing major-account relationships, service them in an exceptional way, and leverage that investment of an additional year in fleshing out the greater part of the visionary's plan. This would not only ensure a secured reference base but also begin to create the infrastructure of ancillary products and interfaces needed to turn a discontinuous innovation into the pragmatist's idea of a real-world solution. Unfortunately, there are no extra dollars in these accounts to pay for this year. Indeed, this year of work is far more likely to be necessary just to catch up to the promises made to secure the deal in the first place. So, while there is plenty of good work to do, there is no money here.

Nor can managers find safety through continuing to service just the early market. To be sure, there are still sales opportunities here—other visionaries who can be sold to. But each one is going to have a unique dream, leading to unique demands for customization, which in turn will overtax an already burdened product development group. Moreover, sooner or later in this early market, yet another entrepreneur with a yet more innovative technology, and with a yet better story to tell, will come along. By then you have to be across the chasm and established in the mainstream, or you are out of luck.

There is still more peril. The marketing efforts to date have been funded by investors—either formally, as in the case of venture-funded enterprises, or informally, as is the case with new products developed within larger corporations. These investors have seen some early successes and now expect to see real progress against the business plan's long-term revenue growth objectives. As we now know, seeking this kind of growth during the chasm period is futile. Nonetheless, it is the commitment in the plan (if the commitment had not been made, the funding would not have been available) and the clock is ticking.

Indeed, a truly predatory type of investor—sometimes referred to as a *vulture capitalist*—looks to use the chasm period of struggle and failure as a means to discredit the current management, thereby driving down the equity value in the company, so that in the next round of funding, he or she has an opportunity to secure dominant control of the company, install a new management team, and, worst case, become the owner of a major technology asset, dirt cheap. This is an incredibly destructive exercise during which not only the baby and the bathwater but all human values and winning opportunities are thrown out the window. Nonetheless, it happens.

Even investors with reasonable demands and a supportive attitude, however, can be troubled by the chasm. Under the best-case scenario, you are asking them to rein back their expectations just when it seems most natural to let them fly. There is an underlying feeling that somehow, somewhere, someone has failed. They may be willing to give you the benefit of the doubt for a time, but you don't have any time to waste. You must get into a mainstream market segment soon, establishing long-term relationships with pragmatist buyers, for only through these can you control your own destiny.

## Fighting Your Way into the Mainstream

To enter the mainstream market is an act of aggression. The companies who  have already established relationships with your target customer will resent your intrusion and do everything they can to shut you out. The customers themselves will be suspicious of you as a new and untried player in their marketplace. No one wants your presence. You are an invader.

This is not a time to focus on being nice. As we have already said, the perils of the chasm make this a life-or-death situation for you. You must win entry to the mainstream, despite whatever resistance is posed. So, if we are going to be warlike, we might as well be so explicitly. For guidance, we are going to look back to an event in the first half of this century, the Allied invasion of Normandy on D day, June 6, 1944. To be sure, there are more current examples of military success, but this particular analogy relates to our specific concerns very well.

The comparison is straightforward enough. Our long-term goal is to enter and take control of a mainstream market (Eisenhower's Europe) that is currently dominated by an entrenched competitor (the Axis). For our product to wrest the mainstream market from this competitor, we must assemble an invasion force comprising other products and companies (the Allies). By way of entry into this market, our immediate goal is to transition from an early market base (England) to a strategic target market segment in the mainstream (the beaches at Normandy). Separating us from our goal is the chasm (the English Channel). We are going to cross that chasm as fast as we can with an invasion force focused directly and exclusively on the point of attack (D day). Once we force the competitor out of our targeted niche markets (secure the beachhead), then we will move out to take over additional market segments (districts of France) on the way toward overall market domination (the liberation of Europe).

That's it. That's the strategy. Replicate D day, and win entry to the mainstream. Cross the chasm by targeting a very specific niche market where you can dominate from the outset, force your competitors out of that market niche, and then use it as a base for broader operations. Concentrate an overwhelmingly superior force on a highly focused target. It worked in 1944 for the Allies, and it has worked since for any number of high-tech companies.

The key to the Normandy advantage, what allows the fledgling enterprise to win over pragmatist customers in advance of broader market acceptance, is focusing an overabundance of support into a confined market niche. By simplifying the initial challenge, the enterprise can efficiently develop a solid base of references, collateral, and internal procedures and documentation by virtue of a restricted set of market variables. The efficiency of the marketing process, at this point, is a function of the "boundedness" of the market segment being addressed. The more tightly bound it is, the easier it is to create and introduce messages into it, and the faster these messages travel by word of mouth.

Companies just starting out, as well as any marketing program operating with scarce resources, must operate in tightly bound markets to be competitive. Otherwise their "hot" marketing messages get diffused too early, the chain reaction of word-of-mouth communication dies out, and the sales force is back to selling "cold." This is a classic chasm symptom, as the enterprise leaves behind the niche represented by the early market. It is usually interpreted as a letdown in the sales force or a cooling off in demand when, in fact, it is simply the consequence of trying to expand into too loosely bounded a market.

The D-day strategy prevents this mistake. It has the ability to galvanize an entire enterprise by focusing it on a highly specific goal that is (1) readily achievable and (2) capable of being directly leveraged into long-term success. Most companies fail to cross the chasm because, confronted with the immensity of opportunity represented by a mainstream market, they lose their focus, chasing every opportunity that presents itself, but finding themselves unable to deliver a salable proposition to any true pragmatist buyer. The D-day strategy keeps everyone on point—if we don't take Normandy, we don't have to worry about how we're going to take Paris. And by focusing our entire might onto such a small territory, we greatly increase our odds of immediate success.

Unfortunately, sound as this strategy is, it is counterintuitive to the management of start-up enterprises, and thus, although widely acknowledged in theory, it is rarely put into practice. Here's the more common scenario.

## How to Start a Fire

Starting a fire is a problem that any Boy Scout or Girl Scout can solve. You lay down some bunched-up newspaper, put on some kindling and some logs, and then light the paper. Nothing could be easier. *Trying to cross the chasm without taking a niche market approach is like trying to light a fire without kindling.*

The bunched-up paper represents your promotional budget, and the log, a major market opportunity. No matter how much paper you put under that log, if you don't have any target market segments to act as kindling, sooner or later, the paper will be all used up, and the log still won't be burning. IBM, for example, burned through tens of millions of dollars trying unsuccessfully to light a fire under PC Jr. That was a very expensive lesson in scouting.

This isn't rocket science, but it does represent a kind of discipline. And it is here that high-tech management shows itself most lacking. Most high-tech leaders, when it comes down to making marketing choices, will continue to shy away from making niche commitments, regardless. Like marriage-averse bachelors, they may nod in all the right places and say all the right things, but they will not show up when the wedding bells chime. Why not?

First, let us understand that this is a failure of will, not of understanding. That is, it is not that these leaders need to learn about niche marketing. MBA marketing curricula of the past 25 years have been adamant about the need to segment markets and the advantages gained thereby. No one, therefore, can or does plead ignorance. Instead, the claim is made that, although niche strategy is generally best, we do not have time—or we cannot afford—to implement it now. This is a ruse, of course, the true answer being much simpler: *We do not have, nor are we willing to adopt, any discipline that would ever require us to stop pursuing any sale at any time for any reason.* We are, in other words, not a market-driven company; we are a sales-driven company.

Now, how bad can this really be? I mean, sales are good, right? Surely things can just work themselves out, and we will discover our market, albeit retroactively, led to it by our customers, yes? The true answers to the previous three questions are: (1) disastrous, (2) not always, and (3) never in a million years.

*The consequences of being sales-driven during the chasm period are, to put it simply, fatal.* Here's why: The sole goal of the company during this stage of market development must be to secure a beachhead in a mainstream market—that is, to create a pragmatist customer base that is referenceable, people who can, in turn, provide us access to other mainstream prospects. To capture this reference base, we must ensure that our first set of customers completely satisfy their buying objectives. To do that, we must ensure that the customer gets not just the product but what we will describe in a later chapter as the whole product—the complete set of products and services needed to achieve the desired result. Whenever anything is left out from this set, the solution is incomplete, the selling promise unfulfilled, and the customer unavailable for referencing. Therefore, to secure these much-needed references, which is our prime goal in crossing the chasm, we must commit ourselves to providing, or at least guaranteeing the provision of, the whole product.

Whole product commitments, however, are expensive. Even when we recruit partners and allies to help fulfill them, they require resource-intensive management. And when the support role falls back on us, it often requires the attention of our most key people, the same people who are critical to every other project we have going. Therefore, whole product commitments must be made not only sparingly but strategically—that is, made with a view toward leveraging them over multiple sales. This can only happen if the sales effort is focused on one or two niche markets. More than that, and you burn out your key resources, falter on the quality of your whole product commitment, and prolong your stay in the chasm. To be truly sales-driven is to invite a permanent stay.

For reasons of whole product leverage alone, the sales-driven strategy should be avoided. But its siren lure is so strong that additional ammunition against it is warranted. Consider the following. One of the keys in breaking into a new market is to establish a strong word-of-mouth reputation among buyers. Numerous studies have shown that in the high-tech buying process, word of mouth is the number one source of information buyers reference, both at the beginning of the sales cycle, to establish their "long lists," and at the end, when they are paring down their short ones. Now, for word of mouth to develop in any particular marketplace, there must be a critical mass of informed individuals who meet from time to time and, in exchanging views, reinforce the product's or the company's positioning. That's how word of mouth spreads.

Seeding this communications process is expensive, particularly once you leave the early market, which in general can be reached through the technical press and related media, and make the transition into the mainstream market. Pragmatist buyers, as we have already noted, communicate along industry lines or through professional associations. Chemists talk to other chemists, lawyers to other lawyers, insurance executives to other insurance executives, and so on. Winning over one or two customers in each of 5 or 10 different segments—the consequence of taking a sales-driven approach—will create no word-of-mouth effect. Your customers may try to start a conversation about you, but there will be no one there to reinforce it. By contrast, winning four or five customers in one segment will create the desired effect. Thus, the segment-targeting company can expect word-of-mouth leverage early in its crossing-the-chasm marketing effort, whereas the sales-driven company will get it much later, if at all. *This lack of word of mouth, in turn, makes selling the product that much harder, thereby adding to the cost and the unpredictability of sales.*

Finally, there is a third compelling reason to be niche focused when crossing the chasm, which has to do with the need to achieve market leadership. Pragmatist customers want to buy from market leaders. Their motive is simple:  whole products grow up around the market-leading products and not around the others. That is, there are many more books about how to use Lotus 1-2-3, many more Lotus-compatible programs, many more Lotus templates, many more directly accessible data sources with 1-2-3, than with, say, SuperCalc or Multiplan or even Excel. The existence of this added-value infrastructure not only enriches the value of the product but also simplifies the job of supporting end users. Pragmatists are very much aware of this effect. They do not want to get caught owning a Honeywell instead of an IBM, a Stellar or Ardent instead of a Convex, a National Semiconductor 32032 instead of an Intel 80386. Therefore, they perhaps unconsciously but nonetheless consistently conspire to install some company or product as the market leader and then do everything in their power to keep them there. One of the main reasons they delay their buying decisions at the beginning of a marketplace—thereby creating the chasm effect—is to help them get a fix on who the leader will be. They don't want to back the wrong one.

Now, by definition, when you are crossing the chasm, you are not a market leader. The question is, How can you accelerate achieving that state? This is a matter of simple mathematics. To be the leader in any given market, you need the largest market share—typically over 50 percent at the beginning of a market, although it may end up to be as little as 30 to 35 percent later on. So, take the sales you expect to generate over any given time period—say the next two years—double that number, and that's the size of market you can expect to dominate. Actually, to be precise, that is the *maximum* size of market, because the calculation assumes that all your sales came from a single market segment. So, if we want market leadership early on—and we do, since we know pragmatists tend to buy from market leaders, and our number one marketing goal is to achieve a pragmatist installed base that can be referenced—*the only right strategy is to take a "big fish, small pond" approach.*

Segment. Segment. Segment. One of the other benefits of this approach is that it leads directly to you "owning" a market. That is, you get installed by the pragmatists as the leader, and from then on, they conspire to help keep you there. This means that there are significant barriers to entry for any competitors, regardless of their size or the added features they have in their product. Mainstream customers will, to be sure, complain about your lack of features and insist you upgrade to meet the competition. But, in truth, mainstream customers like to be "owned"— it simplifies their buying decisions, improves the quality and lowers the cost of whole product ownership, and provides security that the vendor is here to stay. They demand attention, but they are on your side. As a result, an owned market can take on some of the characteristics of an annuity—a building block in good times, and a place of refuge in bad—with far more predictable revenues and far lower cost of sales than can otherwise be achieved.

For all these reasons—for whole product leverage, for word-of-mouth effectiveness, and for perceived market leadership—it is critical that, when crossing the chasm, you focus exclusively on achieving a dominant position in one or two narrowly bounded market segments. If you do not commit fully to this goal, the odds are overwhelmingly against your ever arriving in the mainstream market.

## What About Lotus 1-2-3?

In the legal profession it is often said that great cases make bad law. This is also true in the marketing profession. Let me admit from the outset that the marketing of Lotus 1-2-3 did not follow the niche strategy I have been so strongly advocating. It did not take the D-day approach. Instead, it took what one might call the "Evel Knievel approach": it leapt the chasm.

The Evel Knievel approach is a legitimate—if very high risk—strategy for crossing a chasm. And despite my instinctive aversion to major risks—I am a pragmatist at heart when it comes to adopting new marketing strategies—I would argue that this is the right strategy to take in at least one marketing situation—the one where you are convinced you have a "killer app." In high-tech marketing—specifically in the PC industry—the killer application has enjoyed a status comparable to that of the Holy Grail. It is the application that drives forward an entire market, enriching not only its own vendor but countless other vendors as well. Along with Lotus 1-2-3, WordStar and dBase make up the triumvirate of acknowledged killer apps (with various types of electronic mail arguably a candidate for the fourth). They all appeared in the early 1980s, and for several years the PC industry devoted the bulk of its product-marketing energies to discovering the next killer app. Lotus, for one, almost put itself out of existence in pursuit of this quest. Thus far it has not arrived.

The industry has now been waiting for the next killer app long enough for many to suspect that we are waiting for Godot and should abandon this strategy. But if you believed you were onto such a find, and time were of the essence, and if you had few resources in a world dominated by competitors with many resources, the Evel Knievel approach probably would be the best one to take.

If we have not yet found the Holy Grail, though, if we are mere mortals toiling away in earth time, then I think we need to stick to our niching.

## Beyond Niches

Now, having said all that, we need also acknowledge that there is life after niche. Major market dominance ultimately transcends niche, although it continues to renew and extend itself by developing new segments. And this is indeed when the truly large profits are made. It is clearly a postchasm phenomenon, but there is a planning exercise to be done from the outset. Just as the *objective* of D day was to take Normandy beaches but the *goal* was to liberate France, so in our marketing strategy we want to establish a longer-term vision to guide our immediate tactical choices.

The key to moving beyond one's initial target niche is to select *strategic* target market segments to begin with. That is, target a segment that, by virtue of its other connections, creates an entry point into a larger segment. For example, when the Apple II came out, the initial target segment was the K–12 classroom. This in itself was no small segment, but more importantly, it could be leveraged strategically. Apple II's being used in school came to the attention of the parents of schoolchildren, especially those active in the PTA. They saw their children as benefiting from the programs and bought machines for them at home. These machines, in turn, were put to other uses—scheduling after-school sports programs, writing newsletters for clubs and organizations, keeping mailing lists and small databases. These applications, in turn, caught the interest of the home office business owner, and new applications and a new market segment grew up around promotional mailings, invoicing, and simple inventory.

How successful was all this? When Apple spun off Claris to be an independent software operation in 1987, the latter inherited MacPaint, MacWrite, and MacDraw. By this time the Macintosh was a major marketplace success. But by far the biggest revenue-getter in Claris's entire product line was Appleworks—the only product it carried for the Apple II.

How to ensure that one selects a strategic niche for the D-day landing site is the subject of the next chapter. Before moving on to it, however, let's take a look at some highly visible companies who successfully implemented a highly focused approach to crossing the chasm.

## Successful Chasm Crossings

In the paragraphs that follow we will look at successful chasm crossings by Apple, Tandem, Oracle, and Sun. All four fought off entrenched competition to gain major market shares in the mainstream. The first two took what we are going to call an *application niche* approach. That is, they broke into the mainstream through a particular highly specific application function—desktop publishing and automatic teller machines, respectively. The second two broke through by taking what we will call a *thematic niche* approach. They each identified a single, highly specific theme critical to the broad mainstream market they were after—for Oracle, it was portability; for Sun, open systems architecture—and targeted all their marketing efforts toward securing that position. Although this approach differs markedly from the application niche tack, particularly in the way a company directs its sales force, it has a similar focusing effect on the marketing effort, allowing a smaller, relatively unknown company to force larger, established competitors to get out of its way.

### Exploiting an Application Niche: Apple

In 1986, as a vice-president of a small software company, I was charged with selling Macintosh-based software to corporate America. The best that I could determine at that time was that there were only 4 Fortune 500 MIS departments who had Macintosh on their approved buying list. A year later, however, that number had jumped to over 50, and the year after that to perhaps 350. What happened? Desktop publishing.

*Apple used desktop publishing to cross the chasm.* The graphics department within the Fortune 500 was their Omaha Beach. They landed there in full force, armed not only with their own two key products, the Mac and the Laserwriter, but also accompanied by key allies at Adobe (PostScript) and Aldus (Pagemaker). IBM, who had heretofore successfully barricaded Apple out, did not even try to defend this market niche (and, indeed, had nothing to defend it with). Once Apple was in the graphics department, it got immediate exposure to the sales and marketing people, who were heavy consumers of graphic output, and who, until then, had been alienated from computing and IBM, perceiving both to be in the domain of the finance and operations folks. For these salespeople, the key applications went beyond desktop publishing to more generic functions: MacWrite (and later Microsoft Word) for documents, Excel for spreadsheets, and MacDraw for overhead presentations. The output was sharp, and the computer was kind of fun to use. The sales force was charmed. And soon the Macintosh became the third largest player in the mainstream market for personal computers.

This story may well be familiar—it certainly is to anyone who walks into Regis McKenna Inc.—yet there is a component of it that it is still frequently misunderstood. For several years after the success of desktop publishing, Apple was absolutely bent on finding the next "desktop" success market—desktop presentations, desktop communications, desktop video, and so on. And other PC software and hardware companies also fell into desktop mania. Success, as if often does, had confused people. They were mistaking a one-time tactic (desktop publishing) and a one-time successful event (crossing the chasm) for a repeatable mainstream marketing strategy. Not only was it not repeatable, the very idea of repeating it was irrelevant, because after one has crossed the chasm one is already in the mainstream market and does not need to reenter it.

Moreover, the more time you spend with mainstream customers, the more you see how relentlessly they pursue this conspiracy to sustain market leaders. This means that, once a given tactic has gained one company entry to market leadership, the marketplace instinctively closes over that gap to block further entrance. The mainstream, in other words, functions like an organism in its efforts to expel and reject foreign bodies. So trying to copy the last successful pitch is in fact a very bad strategy, the mainstream being especially alert to blocking this new type of intruder.

### Exploiting an Application Niche: Tandem

Let's look at another successful example of crossing the chasm, this one from the minicomputer business. Tandem computers established a very strong early market position in fault-tolerant computing. They had an absolutely extraordinary claim to make—you could pull one of their computer boards out of the computer *while it was running* and the computer would not go down. In the early eighties, when this claim was being made, computers went down a great deal. Tandem's claim, therefore, was something very special, and it caught the attention of the technology enthusiasts and the visionaries. It also caught the attention of IBM, DEC, and H-P, all of whom began to train their sales forces in how to counterattack.

*Tandem was able to cross the chasm nonetheless, by targeting and dominating an emerging market segment, ATMs (automated teller machines).* We now take these ubiquitous devices for granted, but in fact, they are an invention of the past 20 years, and only really became widespread in the last 10. Tandem took this market by storm because the customer service people in the bank found out very quickly that when an ATM goes down, they have a very unhappy customer, one who is liable to switch banks if things don't change quickly. As a result, Tandem got into virtually every major bank as it switched over to using automated tellers. And once inside these banks, Tandem had a beachhead in the mainstream, which it began to exploit by repositioning itself from being the fault-tolerant leader to being the leader in on-line transaction processing (OLTP).

Now, Tandem is not, in fact, the market leader in OLTP. IBM has something like 49 percent market share, and Tandem about 3 percent. But Tandem nonetheless claimed the high ground, because its machines for a long time were dedicated solely to OLTP while many of the IBM mainframes did the bulk of their processing in batch. Tandem, in other words, was able to optimize its performance for OLTP without a lot of the baggage that the batch machines had to carry. Thus it became the first company to come out with a truly high-performance on-line SQL database, for example, and I can recall sitting in a Gartner Group conference in 1987 in which, in front of perhaps five hundred major MIS types, analyst after analyst sung Tandem's praises above those of any other computer maker. Clearly, the company was well into the mainstream.

Now, there is a central lesson to learn here. Did ATMs exploit the full capabilities of a Tandem computer during this period? Not by a long shot. To pick just one example, Tandem had developed a proprietary networking product called Expand that was the finest technology of its day—but it was not leveraged in the ATM application. So, in a very real sense, to "reduce" a Tandem to an ATM machine was to use it only in one dimension—just as to reduce the Macintosh to a desktop publishing machine was to underutilize its capabilities. How hard for the system designer! To have created a five-speed racing car only to discover that, in its primary use, nobody shifts it out of second gear.

*This reduction in scope is key to the chasm-crossing strategy.* Products that are intended to dominate a mainstream market segment need a well-defined application profile. People value refrigerators, televisions, and pocket knives, but they aren't usually interested in combining all three into one machine. That, unfortunately, is what a computer does. It is so programmable that it can be, in a very real sense, all things to all people. That's a selling point to a technology enthusiast or a visionary. It becomes a neutral proposition to a pragmatist (How much am I paying extra for this capability I am not going to use?) and an absolute negative to a conservative (This is all too complicated for me—I just wanted to buy something to keep my soft drinks cold).

A well-defined application profile, by contrast, presents the innovation in its most attractive light—benefits facing forward, as it were. The boundaries of the market segment are defined by a need for this application, so everyone within the market is viewing the product from the same angle. This has two major advantages. First, by concentrating solely on the application, and by building into the product ancillary features to enhance its capabilities in just that one arena, the newly arrived vendor can win over the normally reluctant pragmatist by sheer virtue of his demonstrated commitment to his specific business. Second, you can leverage what one might call the "Christmas tree effect." That is, because everyone is looking at your product from the same angle, you can turn some of the weaker aspects of your product toward the wall, where no one will notice them. This is not an alternative for the early market, where the technology enthusiast will ferret out everything, but it is sound marketing and development strategy now.

What we are seeing here is nothing less than a conversion from a product-driven to a market-driven strategy—or rather, the beginnings of that conversion. The point is, we are making decisions now that are not in the best interest of the product—something we never did before, and indeed, something that is not appropriate as long as we are in the early market. In that market, we champion the product, and let the customer figure out what it is really for. But in the mainstream, which lacks the time and the talent for this sort of visionary work, the priorities reverse. Now we make decisions that are in the best interest of the market, specifically of the immediate market niche we have chosen as our landing site on the other side of the chasm.

Indeed, once you have undertaken this fundamental change of heart—putting market before product—you realize that there are all kinds of creative ways to present and represent products that had never before been in focus. In the next two examples we will see how two companies used this creative opportunity to create niches out of thin air.

### Exploiting a Thematic Niche: Oracle

First, some background. In the early 1980s relational database management systems (RDBMSs) were an emerging market. The mainstream was held by the major hierarchical databases that ran on IBM mainframes, predominantly IBM's DL/1 (also known as IMS/DB) and Cullinet's IDMS. These databases were optimized for transaction processing throughput, not for analysis and report writing. To get even a small variation in a report typically required an MIS project of one to two weeks. As a result, the first generation of report-oriented databases came into being, led by Information Builder's Focus and Mathematica's Ramis. These were "shadow databases"—loaded up every night from the hierarchical databases that had been updated during the day.

This technology worked reasonably well but was clearly a kludge. What people wanted was one database to serve both roles. That was the promise of the *relational database*. In its early incarnations, however, it simply was too slow to serve as a production database and found its early market acceptance among more research-oriented applications. Interestingly enough, these applications typically were not run on the IBM mainframe but rather on a departmental minicomputer, most often DEC VAXs, and thus the relational database world grew up around DEC and the early Unix machines.

In this world there were two dominant relational database software companies, Oracle and Ingres (then called Relational Technology Inc.), with strong positions in the DEC market, and Informix coming on as a strong third with its developing leadership position in the Unix marketplace. By the end of the decade, however, there was one and only one dominant vendor—Oracle. What happened?

*In the mid-1980s Oracle crossed the chasm by focusing on a single compelling theme—portability across incompatible hardware platforms.* This was a relatively new concern. It used to be that companies could standardize operations around the systems of a single vendor. One of the consequences of pursuing the benefits of downsizing, however, was accepting that the days of a single vendor architecture were gone. No longer could one be an IBM shop, or a DEC shop, or an H-P shop. One was going to have to become all three.

To put further pressure on this issue, management was making it clear that it would no longer accept the notion that data on these incompatible machines could be left unintegrated. The new strategic applications called for the MRP system running on the IBM mainframe to interact with the information in the shop-floor system running on the H-P 3000, which in turn was to get its bill of materials from the design system running on the DEC VAX. The implications for MIS productivity—already beleaguered by a backlog of applications requests—was daunting indeed.

Into this environment Oracle introduced an astounding proposition: there would be one database—Oracle—running on all platforms. This meant that a program written for one platform could easily be ported to another, thereby saving considerable time and effort. It also meant, with networking and gateway products also provided by Oracle, that a program on one computer could call up data on another, incompatible computer, without its programmer having to know anything about that other computer. To give credibility to this proposition, Oracle began porting its product to every platform in sight. To give further credibility, it standardized its database access language on IBM's SQL. This proposition with this evidence won Oracle its beachhead in the mainstream market.

Today, all this seems like obvious good marketing. At the time, it was highly counterintuitive. In the first place, SQL was not the optimal database access language by any means—and why would you ever standardize on a competitor's language anyway? In the second place, most of the platforms that Oracle was porting to were not going to return adequate sales to warrant the effort—so why do it? The bulk of the real business was still just on DEC VAXs, although Unix platforms were beginning to emerge as an interesting segment—why not stick to your knitting? The answer, in retrospect, was that Oracle was sticking to its knitting—people just didn't know what it was that was being knit.

Certainly, Ingres didn't know. It made the same mistake that Apple made when the latter got caught up in pursuing new desktop markets. That is, it mistook a one-time, crossing-the-chasm tactic for an ongoing market need, and it tried, belatedly, to get onto the portability bandwagon instead of finding some other need to exploit to its own advantage. This, again, looked like a reasonable idea at the time but turned out to be a disastrous strategy. Here's why.

The truth was, portability wasn't all that useful. A lot of systems stayed put right where they were written, and the opportunity to leverage people's skills across hardware platforms was more the exception than the rule. Moreover, the gateways and the network access modules that allowed traffic across multiple systems did not, and have not to date, really emerged as viable market products. So there was no major demand backlogging for this kind of service. The market need, in other words, was not to perform on this promise of cross-platform integration but simply to deliver an assurance that this option had not been closed out. Oracle had provided that assurance. In fact, it had made it a key component of the evaluation agenda.

What all this meant was that Ingres had to chase Oracle, just to catch up to what the latter had established as the evaluation criteria, to no real benefit either to Ingres or, more importantly, to its customers. The company developed and executed a strong porting program, the fruits of which are a porting manual for use by hardware vendors that is second to none. Unfortunately, this was a resource that was not spent differentiating Ingres's products to their advantage. Thus, despite a very strong development tools capability, despite a database engine that was at least as good, if not better than, Oracle's, the company languished in the latter's shadow, ultimately to be acquired by its largest VAR, Ask Computer Systems. Recently Ingres finally announced a major reduction in the number of platforms it will continue to support—the final step in recognizing that portability was more a marketing proposition than a deliverable product.

Getting back to Oracle, though, in what sense was all this really niche marketing? In fact, Oracle attacked the Fortune 500 on a relatively horizontal basis. The strategy was never, in other words, to specialize in application niches. Doesn't this directly contradict the D-day analogy?

The answer is, Not if you look at the strategy in the right dimension. If we go back to our three key reasons for niche marketing—simplified whole product requirements, strong word of mouth, and early ability to achieve market leadership—we can see that Oracle's thematic niche strategy met all three goals.

Begin with simplified whole product requirements. Oracle did not have a particularly elegant product heading into this competition. RTI had much better application development tools, and Informix had a far superior 4GL (fourth-generation language). All Oracle's product really offered was availability on a wide variety of hardware platforms and an industry-standard SQL language interface. As far as Fortune 500 MIS shops were concerned, however, that was the whole product. You put up the database, you allocated the disk space, you wrote the SQL calls into your Cobol programs, and it worked. Fourth-generation languages and great application development tools were fine for the technology enthusiasts and the visionaries, but for the pragmatists they were not yet in the mainstream, and thus did not provide a compelling reason to buy. The notion of a portable SQL, on the other hand, did. So, by focusing specifically on that proposition, Oracle successfully and drastically simplified its whole product requirements.

Let's turn now to the word-of-mouth campaign. Here, I think, Oracle was able to leverage two trends in the market that were a consequence of its environment more than of any effort the company itself put out. The first of these was the explosion of Digital Equipment Corporation onto the Fortune 500 scene. DEC was crossing its own chasm at the time. Although its application in departmental management and office automation and its price/performance proposition were compelling, MIS departments were nervous about a whole new set of systems software infrastructure. Oracle's compatibility with IBM standard SQL became a key selling point of the DEC sales force for reducing this tension. It was a claim that was easy to say, never got challenged, and got the sales cycle past that sticking point. So it spread like wildfire.

Now the word-of-mouth community in which it spread consisted of those organizations entertaining the use of DEC equipment in what were traditionally IBM mainframe applications. That community was small enough initially, and the DEC sales force was unanimous enough in its support of Oracle's portability claim, and the claim itself was sufficiently unique, that Oracle was able to achieve word-of-mouth leadership despite its relatively small size at the time.

If this does not seem like a minor miracle, consider how many other database vendors were listed in the Datapro buyers' guides at this time. How many people today are familiar with Adabas from Software AG, Total and Supra from Cincom, IDMS-R from Cullinet, or Datacom-DB from ADR? All these products and companies were far more established than Oracle when the latter began its descent into the chasm. But, with the exception of IDMS, none were able to capture significant word-of-mouth support in the mainstream market. Why? Because none had a unique focus, a thematic niche to exploit.

The third area of concern that drives niche-marketing tactics when crossing the chasm is the need to develop market leadership perception early on in the process. The easiest way to do this, as has already been laid out, is by focusing on a true market segment—something Oracle did not do. What it was able to do instead, however, was to leverage a "virtual market segment," consisting of IBM-oriented Fortune 500 companies who were in the process of committing to DEC systems. Nobody set out to define that segment, but as DEC's value began to surface, this segment did in fact emerge. MIS people within the segment sought each other out to discuss how to cope with the new world, and Oracle quickly became adopted as the market-leader-to-be within this group. Once that happened, the classic conspiracy dynamics propelled Oracle further and further ahead of its closest competitor, Ingres.

So to recap, is this driving focus on a single, compelling theme niche marketing or not? It solves the same problem set that application niche marketing solves. It requires the same intensity of focus and willingness to commit to a restricted strategy that niche marketing requires. But it certainly does not

have the same "look and feel." So before we give a final answer
to this question, let us look at a second example.

### Exploiting a Thematic Niche:  Sun

*What Oracle was able to accomplish through portability, Sun was able
to do through open systems architecture.* That is, it focused all its
marketing resources on gaining ownership of a proposition
that, once again, seems obvious today, but at the time was
startlingly counterintuitive. And by so doing, it successfully
crossed the chasm to gain entry to its mainstream marketplace.

It is hard to convey how wrong the open systems idea sounded inside marketing planning meetings while, at the same time,
it sounded so right in the marketplace. Basically, the whole theory of computer system marketing prior to Sun was based on
creating a differentiating advantage via a proprietary system,
winning customers on the basis of that advantage, and then
using the proprietary architecture to lock those customers in
and competitors out. Every successful system vendor had followed this path. The only one that had not—the IBM PC—had
seen its market taken away by cheap clones (a result so painful
that even after the correctness of open systems marketing had
been established, the IBM group still reverted to a proprietary
bus architecture for its PS/2 line, giving itself the double wham-
my of losing on the open systems proposition from both ends).

At the time Sun launched its open systems strategy, its primary competitor in the workstation marketplace was Apollo
Computer Systems, as entrenched in the mainstream of this
market as any competitor could be, having fought to win the
number one spot from DEC, IBM and H-P. Sun had a great
price/performance proposition, but Apollo had the dominant
position in the marketplace, including, for example, an exclusive relationship with the number-one CAD vendor, Mentor
Graphics. With these kinds of advantages, you do not expect to
be dislodged by any new up-start (which is the way the mainstream likes to spell *start-up*). Again, what happened?

When Sun announced and preached open systems from day
one, it did so in a world that had changed subtly from the days
when proprietary solutions reigned supreme. The market, having suffered the exploitation of being locked in, was ready to be

wooed. It was a little bit like the sixties when, from out of nowhere, a populist movement sprang to the forefront and took control of the national agenda. In the case of Sun, the forum for this protest movement was the standards organizations.

Again, this strategy takes on a highly improbable cast. Standards organizations prior to the mid-1980s were where you sent old engineers to die. They were the rest homes for late adopters, places where innovation was banned and where procedures and bureaucracy held sway. Proprietary vendors gave lip service to their efforts while ensuring that nothing of note could really come out of them. Unix was a joke that kept on living in universities but would never make it in the "real world."

But then the world started to shift. General Motors started something with the MAP initiative, a standard for communications and data protocols on its factory floors. GM demanded of and received compliance from its suppliers to adhere to this standard. This was a major demonstration of customer clout.

MAP didn't directly affect the workstation marketplace to any great degree, but it showed that, with the right kind of muscle behind it, a standards movement could do something. And there was muscle falling in behind the opens systems movement. The federal government, for one, had made Unix a standard, which created a ready-made market too big for any vendor to ignore. The European community got behind the open systems movement. A standard networking architecture (OSI's ISO model) emerged, and on the LAN side, TCP/IP came to the fore.

Finally, there was the de facto example of the IBM PC. Although it was never set up with a standards movement in mind, what grew out of its phenomenal acceptance was one of the most rigorous industry standards of all time. Remember when PC makers wanted to claim that their products were 99 percent IBM compatible? All those vendors were driven from the market. What remained was the first ever binary-compatible platform available at a wide variety of price points from a range of vendors. As a direct result of this, customers experienced dramatic improvements in lowered costs, in higher quality, and in simplification of support tasks. This was something they really liked. And thus the stage was set, although few of us knew it at the time, for Sun.

When Sun entered the workstation market, it brought to the table a staggeringly strong price/performance proposition. But in order to capitalize on this proposition, it had to develop an enormous amount of product infrastructure very, very quickly. That is, it needed to enter the market with state-of-the-art semiconductor chips, workstation systems, application software, a full complement of peripheral devices, and powerful local area networking—*already in place!* That, after all, was what its competitor Apollo was offering. But this is an impossible task for any one company to accomplish, and even more daunting for a start-up.

Sun solved this problem by opening its entire system to other vendors' participation. Sun licensed the architecture of the SPARC microprocessor—*Sun's single greatest proprietary advantage*—to Fujitsu and to Cypress Semiconductor, who could make it and sell it to anyone who would buy. It embraced as its operating system AT&T's Unix and led the movement to standardization around System V. It put its networking solution, NFS, into the public domain. It endorsed all three of the major relational databases—Oracle, Ingres, and Sybase—and refrained (unlike DEC, IBM, or H-P) from creating its own database to compete against them. And by doing all this, it expanded the sum total of resources devoted to developing product infrastructure for SPARC-oriented systems development to something that far exceeded not just its own abilities but Apollo's as well.

Just as important from a competitive point of view, Sun reset the evaluation agenda. Open systems architecture went from being a great development strategy to a great marketing strategy. Now Sun's competitors were the ones on the defensive, for no matter how good Apollo's Domain networking solution was (and it was very good indeed), it was not *open*. DEC's VMS had been the standard engineering operating system (and a very fine one indeed), but it was not *open*. If *desktop* became the Mac's code word of choice, *open* became Sun's.

And, once again, everyone tried to get on the bandwagon. Today, you cannot buy anything that doesn't claim to be open. (It makes a good contrarian think seriously about going out and championing a closed system.) The key point, however, is that only Sun gets the real mileage here—everyone else is chasing *it*, to catch up to its agenda, spending resources that otherwise would be devoted to their own strategy for differentiation. As for the others, the mainstream has already closed over the "open" entry point, and all others attempting to gain acceptance under that umbrella are quickly tagged as also-rans.

So, here again we see that by focusing on a counterintuitive theme, one that sets it apart from the other vendors in its marketplace, a company can achieve the same D-day focus that comes with application niche marketing. To verify this, let us apply the niche marketing tests to Sun's effort. By embracing the open systems thematic niche approach, did Sun simplify its whole product burden? Yes, dramatically, by involving so many third parties so early in the product's life cycle. Was Sun able to create a word-of-mouth effect, and if so, among whom? Yes, it was, and again the dynamics are intriguing, because they may be more circumstantial than planned. Many of Sun's key markets—CAD, CASE, and the U.S. Defense Department, to name three—are heavily populated with technology enthusiasts. As a result, Sun was able to leverage early market word of mouth into mainstream market word of mouth largely because of this overlap. It would not have worked if Sun's mainstream customers had been in office automation, but they were not, and the tactic did work.

Finally, was Sun able to achieve perceived market leadership far earlier than their position relative to Apollo would normally have warranted? Absolutely, and this leadership was a direct function of its ability to champion the open systems agenda like no other company before it. This tagged Sun as the market-leader-to-be even when Apollo had more market share, and thereby allowed it to activate the mainstream market's conspiratorial dynamics in its favor.

So, is this or is this not niche marketing? I think the final answer has to be, who cares? If you can solve the crossing-the-chasm problem set, that is the issue. The solution will require a level of commitment and a narrowing of focus that few companies can muster. The object of that focus must be exceptionally well defined, and devotion to it must be absolute. That is what each of our successful companies demonstrated—a seemingly unwarranted, counterintuitive commitment to an overly restricted proposition that ultimately resulted in mainstream-market acceptance.

## Application Niches or Thematic Niches?

What Apple, Tandem, Oracle, and Sun had in common in their paths to successful penetration of mainstream markets was a single point of focus for their entire company during the crossing-the-chasm period. Two took the application niche route, two the thematic approach. We have seen the upside of both. Before choosing, we had better look at their downside as well.

In the case of Apple and Tandem, who took application niche approaches, this point of focus was discovered empirically, and in each case the people inside the company had mixed feelings about how strongly to support it. This is, perhaps, the biggest drawback of the application niche strategy. The people who design the machine resent its being "downgraded" into an application-specific role, and the people who run the company are afraid of becoming typecast into a narrow niche from which they can never escape. The truth is, neither of these concerns has much long-term merit, but both create powerful resistance at the time, and this in itself can derail an otherwise sound strategy.

Another weakness of the application niche attack is its dependency, over the short term, on the health of its target market. For companies that targeted the semiconductor industry in 1984, for example, this tactic was disastrous. At the time, I was working for a software start-up called Enhansys. A hiatus in PC industry growth (it was going through its own chasm period) put the semiconductor companies into recession that year, and I, having spent a year and a half developing five major accounts, and having confidently agreed in January to bring in a million dollars in 1984 revenues (I already had verbal commitments for more than this), was forecasting goose eggs by April. This is a lesson one does not unlearn.

The alternative tactic of exploiting a thematic niche does not normally involve these problems. It is, first of all, an explicitly intended strategy led from the top (Larry Ellison of Oracle and Scott McNealy of Sun are the key spokespersons for their respective companies), so the tactic does not come as a surprise to people inside the company. Therefore, it does not elicit the same kinds of internal resistance. Second, not being application niche dependent, it does not involve the risk of an industry-specific recession. But it too has its weaknesses.

One weakness is that, at the end of the chasm period, you do not "own" any particular customer base. That is, there is no particular, well-defined, well-bounded application segment that is especially committed to you. This is a drawback, because as you move through the Technology Adoption Life Cycle, beyond the pragmatist group and into the conservative group, there is an increasing application-oriented tendency in customers' communications and buying behavior. If you can own a segment at this point, it will be incredibly loyal to you, as the legal profession has been to Wang, for example, or the publishers to Aldus. Of course, there is nothing to prevent an Oracle or a Sun from initiating application-focused programs once they have crossed over into the mainstream, and in fact, that is what they have both done. But this is not the same, and these programs do not engender the same level of customer loyalty.

Another potential weakness of the thematic niche approach is its vulnerability to being coopted by an established competitor making a quick response. Oracle used to do this to Ingres all the time. Whenever Ingres tried to take the high ground, whether in development tools, database engine capabilities, or multivendor database interconnectivity, Oracle would announce a comparable set of products. Despite having some exceptional technology to offer, Ingres was never able to distance itself from Oracle sufficiently to "own" a differentiating theme.

Perhaps the greatest weakness in thematic niche strategy, however, is simply that it is not always available to be chosen. It requires that there be (1) an identifiable theme in the target marketplace around which your company and its products can be rallied in a natural way, (2) a charismatic spokesperson who can champion this theme effectively, and (3) a sales and marketing management team willing to gamble its entire communications strategy on a single throw of the dice. Having all three come together at the same time simply is not common. And if they have not come together, you cannot force something to happen. For this reason, from here on, we will focus our D-day tactics exclusively on taking an application niche approach to crossing the chasm.

Unlike the thematic approach, the application approach is always available and can be executed successfully under a wide variety of circumstances by virtue of mere competence, provided you are willing to do your homework diligently. As it is laid out in the following chapters, the strategy is optimized for probability of success and stability of results. Both of these gains are achieved, to some degree, through sacrifices in the speed and scope of revenue growth. The object, in other words, is to build a reasonably successful company—not a meteor—taking reasonable risks and making reasonable demands along the way. This has not been the Silicon Valley way. But judging from the current casualty rate, it probably needs to be.

Within the overall context set by the application niche approach to achieving a D-day focus, the first task of the marketing professional is to target the point of attack. And that is the subject of our next chapter.

# 4

# Target the Point of Attack

I don't know who said it—it sounds like the sort of thing that gets attributed to Yogi Berra or to his mentor, Casey Stengel—but in any event, when it comes to crossing the chasm, this saying absolutely holds true: "If you don't know where you are going, you probably aren't going to get there."

The fundamental principle for crossing the chasm is to target a specific niche market as your point of attack and focus all your resources on achieving the dominant leadership position in that segment. In one sense, this is a straightforward market-entry problem, to which the correct approach is well known. First you divide up the universe of possible customers into market segments. Then you evaluate each segment for its attractiveness. After the targets get narrowed down to a very small number, the "finalists," then you develop estimates of such factors as the market niches' size, their accessibility to distribution, and the degree to which they are well defended by competitors. Then you pick one and go after it. What's so hard?

The empirical answer here is, I don't know, but nobody seems to do it very well. That is, it is extremely rare that people come to Regis McKenna Inc. with a market segmentation strategy already in hand, and when they do have one, it is usually

90

not one they are very confident about. Now, these are smart people, and a lot of them have been to business school, and they know all about market segmentation—so it is not for lack of intellect or knowledge that their market segmentation strategies suffer. Rather, they suffer from a built-in hesitancy and lack of confidence related to the paralyzing effects of having to make a *high-risk, low-data decision*.

## A High-Risk, Low-Data Decision

Think about it. We already know that crossing the chasm is a high-risk endeavor, the effort of an unknown and unproven invasion force marching into the camp of some fierce and established competitor. We are either going to get it right, or we are going to lose a substantial portion, perhaps even all, of the equity value in our venture. In sum, there's a lot riding on this kind of decision, and severe punishment for making it badly.

Now, with that in mind, think about having to make what may be the most important marketing decision in the history of your enterprise *with little or no useful hard information*. For since we are trying to pick a target market segment that we have not yet penetrated to any great extent, by definition we also lack experience in that arena. Moreover, since we are introducing a discontinuous innovation into that market, no one has any direct experience with which to predict what will happen. The market we will enter, by definition, will not have experienced our type of product before. And the people who have experienced our product before, the visionaries, are so different in psychographic profile from our new target customers—the pragmatists—that we must be very careful about extrapolating our results to date. We are, in other words, in a high-risk, low-data state.

If you now turn to the established case studies in market segmentation, like as not you will discover they will be based on work done on market share problems *in existing markets*—in other words, work done in situations where there is already a reasonable amount of data to work with. There are precious few paradigms for how to proceed when you cannot examine market share data, indeed cannot even conduct an informed interview with an existing customer of the type you are now seeking to win over. In short, you are on your own.

Now, the biggest mistake one can make in this state is to turn to numeric information as a source of refuge or reassurance. We all know about lies, damned lies, and statistics, but for numeric marketing data we need to open up a whole new class of prevarication. This stuff is like sausage—your appetite for it lessens considerably once you know how it is made. In particular, the kind of market-size forecasts that come out of even the most highly respected firms—the ones that get quoted in the press as showing the bright and promising future for some new technology or product—are, by necessity, rooted in multiple assumptions. Each of these assumptions has enormous impact on the resulting projection, each represents an experienced but nonetheless arbitrary judgment of a particular market analyst, and all are typically well documented in the report, but also typically ignored by anyone who quotes from it. And once a number gets quoted in the press, then God help us—because it has become *real*. You know it is real because pretty soon you see new numbers cropping up, with claims for their legitimacy based on their being derivations of these other "established" numbers.

As you can see, this whole thing is a house of cards. In some contexts, it even has some uses, particularly where financial managers must deal on a macro level with high-tech markets. But it is absolute folly to use such numbers for developing crossing-the-chasm marketing strategies. That would be like using a map of the world to find your way from the Newark airport to the World Trade Center.

And yet, that is what some people try to do. As soon as the numbers get put in a chart—or better yet, a graph—as soon as they thus become blessed with some specious authenticity, they become the drivers in high-risk, low-data situations because these people are so anxious to have data. That's when you hear them saying things like, "It will be a billion-dollar market in 1995. If we only get 5 percent of that market . . ." When you hear that sort of stuff, exit gracefully, holding onto your wallet.

Now, most of the people who come to RMI are more sophisticated than this. They know the numbers do not provide the answers they need. But that doesn't mean they feel any better about having to make a high-risk, low-data decision—which means, in effect, they are stymied. It is our job to get them out of this semiparalyzed state and back into action.

The only proper response to this situation is to acknowledge the lack of data as a condition of the process. To be sure, you can fight back against this ignorance by gathering highly focused data yourself. But you cannot expect to transform a low-data situation into a high-data situation quickly. And given that you must act quickly, you need to approach the decision from a different vantage point. You need to understand that *informed intuition*, rather than *analytical reason*, is the most trustworthy decision-making tool to use.

## Informed Intuition

Despite our culture's anxiety about relying on nonverbal processes, there are situations in which it is simply more effective to substitute right-brained tactics for left-brained ones. Ask any great athlete, or artist, or charismatic leader—ask any great decision maker. All of them describe a similar process, in which analytical and rational means are used extensively both in preparation for and in review of a central moment of performance. But in the moment itself, the actual decisions are made intuitively. The question is, How can we use this testimony to our advantage in crossing the chasm in a reasonable and predictable way?

The key is to understand how intuition—specifically, *informed intuition*—actually works. Unlike numerical analysis, it does not rely on processing a statistically significant sample of data in order to achieve a given level of confidence. Rather, it involves conclusions based on isolating a few high-quality images—really, data fragments—that it takes to be archetypes of a broader and more complex reality. These images simply stand out from the swarm of mental material that rattles around in our heads. They are the ones that are memorable. So the first rule of working with an image is: If you can't remember it, don't try, because it's not worth it. Or, to put this in the positive form: Only work with memorable images.

Now, just as in literature, where memorable characters like Hamlet or Heathcliff or even Dirty Harry stand out and become symbols for a larger segment of humanity, so in marketing can whole target-customer populations become imagined as teeny-boppers, yuppies, pickups and gun racks, or the man in the gray flannel suit. These are all just images—stand-ins for a greater reality—picked out from a much larger set of candidate images on the grounds that they really "click" with the sum total of an informed person's experience. These were, in short, the memorable ones.

Let us call these images *characterizations*. As such, they represent characteristic market behaviors. Teenyboppers, for example, can be expected to shop at a mall, emulate a rock star, seek peer approval, and resist parental restrictions—all of which imply that certain marketing tactics will be more successful than others in winning over their dollars. Now, *visionaries*, *pragmatists*, and *conservatives* represent a set of images analogous to teenybopper, yuppie, and so on—albeit at a higher level of abstraction. For each of these labels also represents characteristic market behaviors—specifically, in relation to adopting a discontinuous innovation—from which we can predict the success or failure of marketing tactics. The problem is, they are too abstract. They need to become more concrete, more target-market specific. That is the function of *target-customer characterization*.

## Target-Customer Characterization: The Use of Scenarios

First, please note that we are not focusing here on target-*market* characterization. The place most crossing-the-chasm marketing segmentation efforts get into trouble is at the beginning, when they focus on a target market or target segment instead of on a *target customer*.

Markets are impersonal, abstract things: the personal computer market, the one-megabit RAM market, the office automation market, and so on. Neither the names nor the descriptions of markets evoke any memorable images—they do not elicit the cooperation of one's intuitive faculties. We need to work with something that gives more clues about how to proceed. We need something that feels a lot more like real people. However, since we do not have real live customers as yet, we are just going to have to make them up. Then, once we have their images in mind, we can let them guide us to developing a truly responsive approach to their needs.

Target-customer characterization is a formal process for making up these images, getting them out of individual heads and in front of a marketing decision-making group. The idea is to create as many characterizations as possible, one for each different type of customer and application for the product. (It turns out that, as these start to accumulate, they begin to resemble one another so that, somewhere between 20 and 50, you realize you are just repeating the same formulas with minor tweaks, and that in fact you have outlined 8 to 10 distinct alternatives.) Once we have built a basic library of possible target-customer profiles, we can then apply techniques to reduce these "data" into a prioritized list of desirable target market segment opportunities. The quotation marks around *data* are key, of course, because we are still operating in a low-data situation. We just have a better set of *material* to work with.

## Pen-Based Laptops:  An Illustrative Example

For the purposes of an illustrative example, let us consider how we might market a tablet-style portable computer, with a touch-screen interface and a stylus. These pen-based laptop-style PCs should be coming on the market around the time this book first hits the shelves. They are certainly getting a lot of advanced press as this manuscript is being completed. Their claim to fame is that you can interact with the computer as if you were using a pen and paper—no keyboard is needed.

Now, let us suppose in the first year or so that pen-based laptop computers win over an early market of technology enthusiasts and visionaries who value in them their unique capability to be used in interpersonal situations, where a keyboard is too intrusive, and away from a desktop, where there is no place to put a keyboard. The Gallup organization becomes an early adopter for use in their polling, and Ford Motor Company makes them a key component of their new quality-reporting system. Now it is time to go after the mainstream market, taking market share away from the traditional keyboard-based laptops. Where would you begin?

This is a classic case of, "So many segments, so little time"— exactly the sort of thing that target-customer scenarios are best for. A representative format for any given scenario is illustrated in the following section. A finished scenario typically would run a page or two at the most—they get much briefer once the organization adopts its own shorthand. As you will see from the example, this is a highly tactical exercise in microcosm, but it has major implications for how marketing strategy is set overall. So as we work through the example, we will also keep an eye out for the broader implications.

## Sample Scenario

1. *Personal profile and job description.* Describe the person (name, age, and so on), the job he or she has, and the type of company worked for, with a view toward displaying both the company's and the person's goals and values. For example:

> Jerome is a 32 year old account executive with Splashi & Splashi, a leading sportswear manufacturer, located in La Jolla, California. He is responsible for placing their new line of Plastique swimwear in sporting goods stores and upscale boutiques, and his territory is northern California. The line is very pricey, and Jerome wants to maintain an upscale, professional image. At the same time, the nature of the sale requires that he able to cite large amounts of highly detailed information while present with the customer at the customer's site. If Jerome carries this information around in a binder, it makes him look like a "bag carrier." But if he doesn't, it slows up the sales cycle considerably—indeed, risks losing the sale.

The idea behind the personal profile is to create an image of a specific target customer in order to stimulate creative imagination of the marketing and R&D teams. Initially these should be highly specific—that's what gets people thinking. Don't worry at this point that you are being too focused. After the team works through a dozen or so, they begin to look more and more alike, and you can move to more general categories—but not until then.

The broader implication here is that most marketing teams fail right at the beginning. They do not personalize their image of the market but go straight to some set of abstractions or numbers. What gets lost in the process, as we shall see, is any clear insight into the compelling reason to buy.

2. *Technical resources.* List the technical resources available to the target customer as they pertain to the use of the new product. For example:

> Jerome himself has never used a personal computer, but he works in an office that is equipped with several IBM PCs, which are connected to the main computer at the head office. The PCs are equipped with modems and printers. The office also has a fax machine.

This is the other half of the target-customer profile—the unseen layer of technical support upon which the scenario's value propositions must rely. It is particularly important information for high-tech marketing decisions, since a lot of technical solutions imply the existence of other technical products for them to work properly. We need to know how much infrastructure we expect to be in place to support this would-be end user. Not only will this impact the marketing planning process, but it will also serve as a key qualifier during the sales cycle.

The broader implications of what technical infrastructure is, or is not, already in place will come out in the next chapter when we discuss the concept of "whole product" and its impact on market domination.

3. *A day in the life (before).* Write a brief dramatization of the exact moment in the target customer's life when the new product would be of most benefit. For example:

> Jerome is in the midst of taking an order for the basic line of Plastique swimwear at the Ghirardelli Square Windsurf and Kite Store. He notices that the other

sportswear on display features at lot of fluorescent colors. Plastique is coming out with a new line of fluorescent wear, but Jerome is not sure when or at what price. He pulls out one of his binders full of information, only to have several leaflets spill out onto the floor. As he is bending down to pick these up, his customer looks at her watch. He knows he can find the material, but it is going to take some time. As he is frantically paging through the catalog, she says, "Tell you what, let's just stick with the basic line for now."

This scenario contains the seeds of the value proposition. The idea at this point is to state the target customer's problem or opportunity through concrete imagery, not abstract terms. This is a classic idea-generation technique: just let the "right brain" drive at this point; the left brain can come along later and sift through the materials to separate the wheat from the chaff.

4. *The dilemma.* State as clearly as possible the problem that is going to motivate the target customer to buy the proposed product. For example:

> To sell the maximum amount of high-margin product line, Jerome must maintain a highly professional image and be able to reference large amounts of detailed information at a moment's notice. Jerome's inability to do this more efficiently is costing his company sales and him commissions.

Now it's the left brain's turn. By abstracting the problem, we transform the single target customer into a representative icon for a target market—in this case, *Jerome* becomes *upscale sales professionals who sell out of catalogs.* At the same time we are articulating as bluntly as possible the proposition that creates a compelling reason to buy.

5. *A day in the life (after).* Revise the dramatization, this time illustrating how the proposed product solves the dilemma, and how that reinforces the values of the target customer and his or her company. For example:

> Having noticed the interest in fluorescent colors, Jerome touches the button on the screen of his pen-based laptop that says "reference materials." This calls up a display of

several icons, and he selects "new products," again just with a touch of the stylus onto the screen. Jerome explains to his customer that inside the computer is information downloaded from the corporate mainframe to the PCs at Jerome's office. From there, it has been downloaded again, by the office secretary, into Jerome's tablet. Two or three more touches narrow the information down to what he wants. His customer, meanwhile, pulls her chair closer to him to see how the system works. When the information comes up on the fluorescent swimwear, she says, "Hey, that's really neat. So how much would that extra line of bathing suits cost?"

This scenario speaks to the central question: What does success look like? This question is more useful than any other in focusing a team effort like crossing the chasm. Ultimately, the scenario of success needs to be abstracted into a value proposition that encompasses a broader set of target customers. But again, it is vital to let the intuitive capabilities of concrete imagery do their work first.

## Processing Characterization Data: The Value Triad

Target-customer characterization is at the core of applying market segmentation strategy to the problem of crossing the chasm. It supplies the "data." Assume that we have spent several weeks building a broad library of target-customer scenarios—say, 50 or so, developed by a cross-functional marketing team of six to eight members. We then submit those scenarios to a one-pass review to consolidate them—weeding out the totally implausible and combining the ones that are essentially the same. The result is a body of material on potential target customers. The model into which that material is fed is the *value triad*.

The value triad is made up of three components: the product itself, the target customer, and the target application. The three taken together make up a value proposition, something one can sell. That is, an eggbeater, a cook, and the baking of a cake combine to create a value proposition called "saving time and energy," which can be used to sell the eggbeater to a cake baker.

# The Value Triad

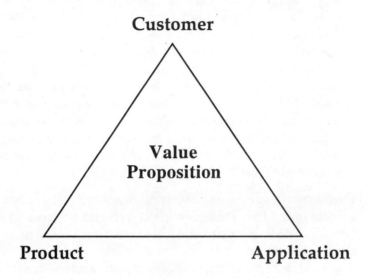

Cake bakers thereby become a segment of the eggbeater market, other segments of which might include waffle makers, soufflé chefs, and instant pudding preparers.

If you change any element in the triad, you change the value proposition. If, for example, the target application is not mixing cake batter but instead is folding in the soufflé's egg whites, then a whole new set of concerns about overmixing comes up, and the eggbeater's value goes down. If the target customer is not a chef but a small child making instant pudding, then the concerns might shift to how safe the product is for small fingers. This, in turn, might lead to a request to modify the product to include plastic-coated blades. And so on.

The value triad is the initial basis for assessing the attractiveness of any target-market opportunity. As potential characterizations are fed into the model, the goal is to tune the outer elements of the triangle—product, customer, and application—so that the inner element, the value proposition, becomes the most compelling possible reason to buy. *Prioritizing opportunities based on compelling reason to buy is the primary tool for identifying the target segment for crossing the chasm.*

In the case of crossing the chasm, there are some additional limits imposed on the prioritizing process. In the first place, we normally cannot make much of a change to *product* at this point. We lack the time, the resources, and the cash flow to do so. The tuning effort, then, is pretty well confined to two variables: customer and application. And second, the customer variable is further limited by being in the pragmatist set, for the whole point of crossing the chasm is to win over this new type of customer.

In its most bare-bones form, the solution universe for this problem is a two-dimensional spreadsheet, the rows of which list all the possible applications, and the columns of which list all possible target customers. At the intersection of any row and column is placed a value proposition rating. This rating is based on the consensus of the marketing team. A reasonable system is based on a one-to-five scale, as follows:

1. Not usable.
2. Usable, but with no obvious benefits.
3. Nice to have—the end user will appreciate these benefits although they are not strategic to the organization sponsoring the purchase.
4. Should have—the end user receives strategic benefits, although these benefits can readily be achieved by other means as well.
5. Must have—the end user receives benefits that are strategic to the sponsoring organization and cannot be achieved by any other reasonably comparable means.

Here is a sample section of such a spreadsheet for a pen-based laptop computer:

### Pen-based Laptop Value Matrix

| Application | Customer | | | |
|---|---|---|---|---|
| | "Jerome" | Product Mgr. | CEO | etc. |
| Selling from catalogs | 4 | 1 | 1 | |
| E-mail or fax while traveling | 2 | 3 | 4 | |
| Meetings notes management | 3 | 4 | 3 | |
| Editing documents away from desk | 2 | 3 | 4 | |

First, please note that there are no 5s in this example—no *must-haves*. This is no accident. Must-haves are as rare as they are desirable. The best we could do was to give four 4s or should-haves—one to Jerome for catalog sales, two to the CEO for keeping in better touch with the office while traveling and for editing documents away from her desk, and one to the product manager for better organizing all his meeting notes. This is not good enough.

## The Compelling Reason to Buy: The Prioritizing Factor

*To cross the chasm you must target a market segment defined around a must-have value proposition.* Only a must-have value proposition is strong enough to overcome the pragmatist's natural aversion to embracing a discontinuous innovation before it has evolved into an accepted market standard. Should-have propositions won't normally cut it, because a pragmatist is too wary of the chaos that discontinuity carries with it.

In the most abstract sense, there are three sources of a must-have condition:

1. *It enables a previously unavailable strategic capability that provides a dramatic competitive advantage in an area of prime operational focus.*

   *Example:* Having its truck drivers use hand-held computers allows the Frito-Lay company to achieve an otherwise unheard-of freshness level in its on-the-shelf inventory.

   This type of benefit has greatest appeal to a *visionary*. It cannot be cost justified except in the most abstract terms. Nonetheless, the visionary is intent on betting her career—as well as a large chunk of her company's resources—on achieving strategic breakthrough. Therefore, she must have some technology upon which to build.

   This is the type of value proposition that builds *early markets*. Enterprises typically come to the chasm having had some considerable success in this domain. Now that it is time to cross the chasm, however, this value proposition must be abandoned in favor of the following one.

2. *It radically improves productivity on an already well-understood critical success factor.*

   *Example:*   On-line, fault-tolerant transaction processing from Tandem computers on the New York Stock Exchange allows for much greater transaction throughput with much higher degrees of reliability and with much faster confirmation to the customer.

   This benefit has greatest appeal to a *pragmatist*. It is typically cost justified in terms of soft dollar savings—that is, getting a better return on resource expenditures than before. The pragmatist is committed to making improvements within established guidelines that have already been laid out, and he must have technology products to bring about those improvements. (At the same time, however, he also must have a much lower risk exposure than the visionary's.)

   This is the value proposition that is fundamental to crossing the chasm. It is less dramatic than the first but has broader appeal—hence, its ability to sustain a mainstream market segment. Note that it, in its turn, is distinct from the following.

3. *It visibly, verifiably, and significantly reduces current total overall operating costs.*

   *Example:* Replacing typewriters with PC-based word processors means cutting headcount by hiring fewer secretaries to support the same number of people. The added cost of the PC-based system is paid back in a matter of months.

   This benefit has the greatest appeal to a *conservative*. It is typically cost justified in terms of hard dollar savings—that is, direct reductions to the expense budget. The conservative is committed to providing only the most standard services—but at the best price. She must have low-cost, high-quality technology products to accomplish this goal.

   This type of benefit is normally inappropriate to crossing the chasm. The chasm's risk factors are still too high for conservatives to take a chance, and there is still too much investment needed in the infrastructure necessary to make the application cost-justifiable in hard dollars in the short term.

So, to recap this point, whenever you are crossing the chasm, seeking to win over a pragmatist-oriented segment of a mainstream marketplace, the compelling reason to buy is always a version of the following value proposition: *Our new product radically improves productivity on an already well-understood critical success factor specific to your business, and there is no existing means by which you can achieve a comparable result.*

For this claim to be credible to a pragmatist, the vendor must demonstrate familiarity with the specifics of the business and demonstrate a product that integrates cleanly with existing systems—no small tasks. They require time spent in the industry, talking and working with the participants, and they require investment in bridging the gaps between the product and the other systems in place. This is both expensive and time-consuming, hence, the need to focus on a specific target market segment.

## Is the End User Always the Target?

There is one final point to make about the target-customer characterization process before tying it into the larger effort of selecting the point of attack. With high-tech products there is not exactly "one" target customer, even within a specific characterization like Jerome's. After all, Jerome is not the buyer of his tablet. That purchase will almost certainly come out of the vice-president of sales's budget. And the vice-president is not likely to buy the tablet unless he has the approval and support of the MIS manager. And the MIS manager is not likely to give that approval unless her technical specialist gives it the green light, and her operations people have determined the impact and feasibility of supporting it. Probably, in fact, someone is going to want to conduct a pilot test, and that, in turn, is likely to involve some early adopters, including the secretary in Jerome's office who is supposed to be able to do all the download magic.

The question then arises: If all of these other people are somehow part of the target customer, don't we have to characterize them as well? The answer is, Not in the same sense. For the purposes of high-tech product marketing, everyone other than the end user can be stereotyped in such roles as economic buyer, MIS, techie, and the like. The reason is that these types do not vary much from market to market—they are inherent in high-tech sales of any kind. *It is almost always the end user and his or her application that give the market its defining characteristics.*

This principle of focusing primarily on characterizing the end user holds true even if the product will be sold exclusively to an OEM or VAR or some other indirect means for reaching the final customer. It is the end user's dilemma and the solution to it that set the value of the product and determine the other elements (products, people, conventions) that have to be taken into account in order to field a successful product offering. Companies who market exclusively to OEMs—semiconductor companies, for example—used to be able to ignore the end user entirely and trust entirely to the vendor to pull all the parts together. Now, however, solutions have become so complex, and the number of elements in any total solution is so great, that system vendors need their chip suppliers to be thinking ahead, trying to anticipate and simplify the challenges of downstream integration problems. The only way to do that consistently is to characterize the end user, feel what it's like to walk in this person's shoes, and envision all that is needed to solve the problem at hand.

## Selecting the Point of Attack

Creating the target customer scenarios, distributing them across a two-dimensional spreadsheet, and ranking their value propositions capture a reasonable universe of alternatives for selecting a point of attack. This is the first "market map" to put up in your war room. The great advantage of this map is that it not only highlights where the 5s are but also can show where there are clusters of 4s. *Should-haves* are great target-market reinforcers. Indeed, once the product has become established in the mainstream, they can become market-building propositions in their own right. So the more of these you can incorporate into the initial marketing attack plan, the better.

Specifically, if you can find either a row or a column on the spreadsheet that is dominated by high ratings, then you are really onto something exciting—either a key target customer wanting many applications or a key target application having many customers. In both cases there is leverage to exploit.

In the case of many applications for a single customer, it is distribution leverage. Having created the relationship with a given customer via one application, you can leverage it to sell in the other applications. There is no need to put in all the work to establish a new channel.

In the case of many customers for a single application, it is whole product leverage. Having built the whole product infrastructure to support the application for one type of customer, you can leverage that investment by using virtually the same product infrastructure in support of another. There is no need to develop a whole new set of supporting interfaces or go out and recruit a whole new set of partners and allies.

So finding clusters of should-haves makes good economic sense from a leverage viewpoint. It should not distract us, however, from the primary selection principle: *Always, at the center of the point of attack, there needs to be a target customer with a truly compelling, must-have reason to buy.*

Once the market map's universe of opportunity has been narrowed down to a reasonably small set of potential points of attack, it is time to bring back traditional market research and analysis techniques. In order to employ these techniques, one must first translate the high priority customer scenarios into market segment definitions. That is, a *Jerome* might be translated as "upscale, catalog-oriented sales forces." The latter phrase must define the set of potential customers such that, with one or two screening questions, market researchers (and later, salespeople) can determine whether the company or person they are talking to is part of the right set. The cleaner and crisper these screeners are, the more confidence you can have in the market research you do, and the more efficient your subsequent lead-generation activity can be.

For each segment that is a truly promising candidate for attack, traditional market research should be done both on a macrolevel—how big the population of Jeromes is, where they are located, how much computer investment has been made in them to date—and on a microlevel, through focus group discussions with qualified Jeromes and/or their bosses—how compelling the value proposition really is, what is going to get in the way of the buying decision, what reassurances are needed, and so on.

The key gatekeeper on the macrolevel is the principle that the targeted segment must be of a sufficient size that all of next year's revenue target can be achieved from selling to a reasonable percentage of customers within this segment. This means that the segment must be large enough and accessible enough to make this a reasonable proposition. If it is not, then it either must be replaced by or supplemented with another segment. In the latter case, the gatekeeping factor shifts to whole product support—how many segments can we realistically serve? There are enormous advantages to keeping this number at one.

The key gatekeeper on the microlevel—the factor that comes under scrutiny in the focus group work—is the strength of the value proposition versus the resistance evoked by the discontinuous innovation. It turns out that even minor changes in how the reason to buy is framed can have a major impact on this balance. It also turns out that major resistance often can be traced back to highly unexpected sources, little things about the product or the way it is distributed that can be readily changed. The key here is to conduct the focus groups so that the respondents get a very concrete picture of what the product, the buying process, and the postsales support are actually going to look like. Only then can they give you the kind of feedback that will make a difference.

Having reviewed the macro- and microlevel market research findings, we are left with the final step of prioritizing our market segments based on everything we know at this point. The result is a list of "hot targets." The commitment at this point is that (1) any one of these is a legitimate point of entry to the mainstream market, and (2) there is not likely to be some other point of entry that we haven't thought of yet.

All subsequent strategy planning is driven off of the hot-target list. Finally, however, one cannot choose the landing site until one has thought through and researched two more sets of key issues. First, how strong an invasion force can we assemble when attacking our target segment? And second, what kind of competition will we face, and how well are we equipped to deal with it? These issues are the subjects of the next two chapters.

## Recap: The Target Market Segment Selection Checklist

We have been saying all along that the material in this chapter and the following three chapters is tactical by nature—that is, made up of relatively specific tasks and exercises that can, and should, be performed recurrently throughout a major enterprise. As a way of recapping this material, at the end of each chapter there will be a checklist of activities, suitable as a means either for managing a group through this process or testing the final output of a group's marketing decision making.

For selecting the target market segment that will serve as the point of entry for crossing the chasm into the mainstream market, the checklist is as follows:

1. Develop a library of target-customer scenarios. Keep adding to it until new additions are nothing more than minor variations on existing scenarios.

2. Analyze these scenarios into customer and application types, building a spreadsheet in which there is a new row for each new customer, and a new column for each new application.

3. Using the scenarios as a guide, rate the compelling reason to buy at each customer-application intersection.

4. Define potential market segments around areas of the spreadsheet where there is a least one must-have rating accompanied by a cluster of should-haves, either in the same row (same customer) or same column (same application).

5. Apply traditional macro- and microlevel market research techniques to evaluating each of these candidate market segments.

6. Use the market research to impose a final priority order on the candidate segments, producing a hot-target list.

# 5

# Assemble the Invasion Force

*"I have always found you get a lot more in this world with a kind
word and a gun than you do with just a kind word."*
—Willie Sutton

Willie is only restating what any military leader will confirm: If
you are committing an act of aggression, you'd better have the
force to back it up. Or, to put this in terms closer to our immediate topic, marketing is *warfare*—not *wordfare*.

Which of us, about to launch an invasion, would prefer a
good set of slogans to a good set of offensive and defensive
weapons? Who would rather buy advertising time on televi-
sion than missiles and munitions? Who would rather publish a
manifesto than have guaranteed treaties with neighboring
countries? Most high-tech executives—that's who.

There is a widespread perception among high-tech executives
that marketing consists primarily of some long-range strategic
thinking (when you can afford to take the time for it) and then a
lot of tactical sales support—with nothing in between. In fact,
marketing's most powerful contribution happens right in
between. It is called *whole product marketing*, a term introduced
earlier, and it is the fundamental basis for assembling the invasion force.

Consider the following scenario. When I was a salesman, I had a dream. The dream was simple. There was a monster bid coming up—with a $5 million minimum—and I had *wired* the request for proposal (RFP). I had, in the words of gamblers everywhere, a *mortal lock* on the thing. The client had met with me for long hours of consultation during which he had bought into every selling argument in favor of my product. He had then constructed the RFP so that only my product could get a 100 percent evaluation. The deal was mine. Then I woke up.

Okay—so that's a fantasy. But a version of that fantasy can be executed in the real world. We might call it *wiring the marketplace*. Again, the concept is simple. For a given target customer and a given application, create a marketplace in which your product is the only reasonable buying proposition. That starts, as we saw in the last chapter, with targeting markets that have a *compelling reason to buy* your product. The next step is to ensure that you have a monopoly over fulfilling that reason to buy.

To secure that monopoly, you need to understand (1) what a *whole product* consists of and (2) how to organize a marketplace to provide a whole product incorporating your company's offering.

## The Whole Product Concept

One of the most useful marketing constructs to become integrated into high-tech marketing in the past few years is the concept of a whole product, an idea described in detail in Theodore Levitt's *The Marketing Imagination,* and one that plays a significant role in Bill Davidow's *Marketing High Technology* . The concept is very straightforward: There is a gap between the marketing promise made to the customer—the compelling value proposition—and the ability of the shipped product to fulfill that promise. For that gap to be overcome, the product must be augmented by a variety of services and ancillary products to become the whole product.

The formal model is diagramed by Levitt as follows:

## The Whole Product Model

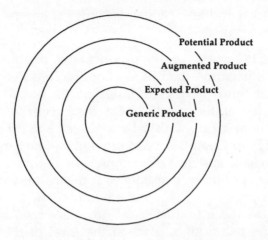

The model identifies four different perceptions of product, as follows:

1. *Generic product:*  This is what is shipped in the box and what is covered by the purchasing contract.

2. *Expected product:*  This is the product that the consumer thought she was buying when she bought the generic product. It is the *minimum* configuration of products and services necessary to have any chance of achieving the buying objective.

   For example, people who are buying personal computers for the first time *expect* to get a monitor with their purchase—how else could you use the computer?—but in fact, in most cases, it is not part of the generic product.

3. *Augmented product:*  This is the product fleshed out to provide the *maximum* chance of achieving the buying objective.

   In the case of a personal computer, this would include a variety of products, such as software, a hard disk drive, and a printer, as well as a variety of services, such as a customer hot line, advanced training, and readily accessible service centers.

4. *Potential product:* This represents the product's room for growth as more and more ancillary products come on the market and as customer-specific enhancements to the system are made.

In the PC world, on the product side, this means having an open architecture—the bus slots on the IBM PC, for example—such that third parties can build add-in boards to extend the capabilities of the product. On the service side, this might involve hooking up with network-based service providers such as Prodigy or CompuServe to provide on-line banking, information services, and home shopping. In the area of supporting customer-specific enhancements, it means providing some sort of development language and environment.

Now, at the introduction of any new type of product, the marketing battle takes place at the level of the generic product—the thing in the center, the product itself. This is the hero in the battle for the *early market.* But as marketplaces develop, as we enter the *mainstream market,* products in the center become more and more alike, and the battle shifts increasingly to the outer circles. To understand how to dominate a mainstream marketplace we need to take a closer look at the significance of what Paul Harvey might call *the rest of the whole product.*

## The Whole Product and the
## Technology Adoption Life Cycle

First, let's look at how the whole product concept relates to crossing the chasm. If we look at the Technology Adoption Life Cycle as a whole, we can generalize that the outer circles of the whole product increase in importance as one moves from left to right. That is, the customers least in need of whole product support are the technology enthusiasts. They are perfectly used to cobbling together bits and pieces of systems and figuring out their own way to a whole product that pleases them. In fact, this is in large part the pleasure they take from technology products—puzzling through ways to integrate an interesting new capability into something they could actually use. Their motto is: Real techies don't need whole products.

For the visionaries, there is no pleasure in pulling together a whole product on their own, but there is an acceptance that, if they are going to be the first in their industry to implement the new system—and thereby gain a strategic advantage over their competitors—then they are going to have to take responsibility for creating the whole product under their own steam. The rise in interest in systems integration is a direct response to increasing visionary interest in information systems as a source of strategic advantage. Systems integrators could just as easily be called whole product providers—that is their commitment to the customer.

So much for the market to the left of the chasm, the early market. To get to the right of the chasm—to cross into the mainstream market—you have to first meet the demands of the pragmatist customers. These customers want the whole product to be readily available from the outset. They like a product such as WordPerfect because there are not only books in every bookstore about how to use it but also seminars for training, the industry's best hot-line support (out of Orem, Utah, no less), and a whole cadre of temporary secretaries already trained on the product. If instead the pragmatists are offered a "great deal" on an alternative product, say Ami from Samna (now from Lotus), they just aren't interested because the whole product isn't there.

The same logic holds for why they prefer IBM's hardware to Tandem's, Lotus 1-2-3 to Wingz, Intel 486–based workstations to MIPS-based RISC workstations, Winchester disks to Flash EEPROM solid state storage, Novell LANs to Banyan, and Oracle to Unify. In every case, there are strong arguments that they are preferring an inferior product—if you look only at the generic product. But in every case, they are preferring the superior product, if you look at the whole product.

*To net this out: Pragmatists evaluate and buy whole products.* The generic product, the product you ship, is a key part of the whole product, make no mistake. But once there are more than one or two comparable products in the marketplace, then investing in additional R&D at the generic level has a decreasing return, whereas there is an increasing return from marketing investments at the levels of the expected, the augmented, or the potential product. How to determine where to target these investments is the role of whole product planning.

## Whole Product Planning

As we have just seen, the whole product model provides a key insight into the chasm phenomenon. The single most important difference between early markets and mainstream markets is that the former are willing to take responsibility for piecing together the whole product (in return for getting a jump on their competition), whereas the latter are not. Failure to recognize this principle has been the downfall of many a high-tech enterprise. Too often companies throw their products into the market as if they were tossing bales of hay off the back of a truck. There is no planning for the whole product—just the hope that their product will be so wonderful that customers will rise up in legions to demand that third parties rally about it. Well, God did divide the Red Sea for Moses.

For those who wish to take a more prudent course, however, whole product planning is the centerpiece for developing a market domination strategy. Pragmatists will hold off committing their support until they see a strong candidate for leadership emerge. Then they will back that candidate forcefully in an effort to squeeze out the other alternatives, thereby bringing about the necessary standardization to ensure good whole product development in their marketplace.

A good generic product is a great asset in this battle, but it is neither a necessary nor a sufficient cause of victory. Oracle did not have the best product when the market standardized on it. What Oracle offered instead was the best case for a viable whole product—SQL standardization plus broad portability across hardware platforms plus an aggressive sales force to drive product into the market quickly. That is what the pragmatists in MIS got behind.

In short, winning the whole product battle means winning the war. And, because perception contributes to that reality, looking like you are winning the whole product battle is a key weapon to winning the war. On the other hand, *pretending* you are winning the whole product battle is a losing tactic—people check up on each other too much in the high-tech marketplace. These distinctions will become critically important in our next chapter, where we deal with *positioning*.

For now, our focus should be on the minimum commitment to whole product needed to cross the chasm. That is defined by that whole product which assures that the target customers can fulfill their compelling reason to buy. To work out how much whole product this is, you only need a simplified version of the whole model:

## The Simplified Whole Product Model

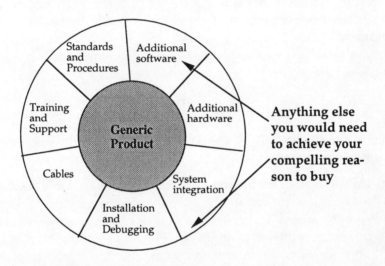

In the simplified model there are only two categories: (1) what we ship and (2) whatever else the customers need in order to achieve their compelling reason to buy. The latter is the *marketing promise* made to win the sale. The contract does not require the company to deliver on this promise—but the *customer relationship* does. Failure to meet this promise in a business-to-business market has extremely serious consequences. As the bulk of purchases in this marketplace are highly reference oriented, such failure can only create negative word of mouth, causing sales productivity to drop dramatically.

Classically, high tech has delivered 80 to 90 percent of a whole product to any number of possible target customers, but

100 percent to few, if any. Anything less than 100 percent, unfortunately, means that the customers either supply the remainder themselves or feel cheated. Significantly less than 100 percent means that the target market simply does not develop as forecast—even if the generic product, the product in the box being shipped, is superior to anything else in its class.

In short, if you wanted to trace disillusionment with high tech's inability to deliver on its promise to its investors and its customers, lack of attention to whole product marketing is the closest thing to a wellspring. This is actually great news—it means that the converse applies as well. By solving the whole product equation for any given set of target customers, high tech has overcome its single greatest obstacle to market development.

Let's look at an example to see how this works out.

## Jerome, Revisited

Let's revisit the "after" scenario for Jerome and his pen-based laptop computer:

> Having noticed the interest in fluorescent colors, Jerome touches the button on the screen of his pen-based laptop that says "reference materials." This calls up a display of several icons, and he selects "new products," again just with a touch of the stylus onto the screen. Jerome explains to his customer that inside the computer is information downloaded from the corporate mainframe to the PCs at Jerome's office. From there, it has been downloaded again, by the office secretary, into Jerome's tablet. Two or three more touches narrow the information down to what he wants. His customer, meanwhile, pulls her chair closer to him to see how the system works. When the information comes up on the fluorescent swimwear, she says, "Hey, that's really neat. So how much would that extra line of bathing suits cost?"

Now, let's analyze that scenario in terms of its implied whole product commitments. There are several:

- *"Touches the button . . . for reference materials."* What button? Who put it there? Clearly this is an application-specific button, so that rules out the hardware manufacturer. It also rules out any shrink-wrapped software supplier as well. That gets us down to an in-house programming effort or a custom effort supplied by a systems integrator or VAR.

That, in turn, implies a software development toolkit, which includes not only a language and a library of sub-routines but also a development environment, debugging tools, and training in their use.

In short, that little button is a very, very expensive item, with all kinds of chances never to get created.

- *"Information downloaded from the corporate mainframe to the PCs."* By whom? This kind of activity implies a number of data extraction utilities, both on the PC and on the mainframe, a lot of corporate agreement about how information will be formatted, and possibly some custom application development by the MIS group at the mainframe.

  In addition, it implies an even higher level of training for Jerome (remember, he was initially described as someone who does not use a PC). Mainframe-based information tends to be astoundingly voluminous and organized in ways that most people would not find easy to comprehend. If you don't train Jerome, then chances are he will not risk making a fool of himself in front of a client and hence not use the system in its intended—and compelling—application.

- *"Downloaded . . .* [from the PC to the] *tablet."* Again, by whom? The phrase implies some additional file format conventions, this time ensuring compatibility between the PC and the tablet computer, as well as some connectivity hardware and software, and a person other than Jerome who is knowledgeable enough to operate the downloading system. And how often does this downloading have to take place to keep Jerome current? And how does Jerome know about new information that has been downloaded?

And so on. *The point is, even a single target-customer profile starts off a chain of issues that any product manager serious about developing a particular market opportunity must pursue to a satisfactory conclusion.*

Now, in the case of an electronic tablet you might imagine a fairly lengthy list of potential target customers and target applications. In addition to salespeople like Jerome, one could imagine:

- Traffic cops writing tickets (now it's become an input device, so the stylus has to do handwriting or at least printing, and it will need a printer if they are going to leave behind a copy of the tickets they write).
- Presentation givers, especially for one-on-one situations (which means it ought to support at least some of the popular presentation software like PowerPoint or Persuasion).
- Personal organizer or Filofax users (which means it needs some kind of forms generation capability, or a third-party supply system making lots of alternative forms—something like the people who support HyperCard with stacks).
- Airline pilots, or anyone else who has to go through checklists on a routine basis (which means an output device to send the record of the check to some larger computer repository).
- Inventory takers (which means some rudimentary numeric data entry pad, along with a good built-in find/search utility).

As even this cursory listing indicates, *every additional new target customer will put additional new demands on the whole product*. That is, the total sum of products and services needed in order to get the desired benefit changes any time you change the value proposition. It soon becomes clear to even the most optimistic product marketing managers that they cannot go after all markets at once, that at minimum they have to sequence and prioritize opportunities, and that each opportunity has very real support costs.

Now, given the need for a whole product in order to fulfill the customer's reason to buy, what is the responsibility of the tablet computer hardware vendor—and specifically of the product marketing manager who has the tablet PC as his product—for seeing that this whole product is in fact delivered? The answer is, it has nothing to do with responsibility, it has to do with marketing success. If you leave your customer's success to chance, you are giving up control over your own success.

Conversely, by thinking through your customer's problems—and solutions—in their entirety, you can define—and work to ensure that the customer gets—the whole product.

At no time is this marketing proposition more true than when crossing the chasm. Prior to the chasm there is some hope that the visionaries will backfill the whole product through their own systems integration efforts. Once the product is established in the mainstream, there is some hope that some third party will see an opportunity for itself to make money fleshing out the whole product. *But while you are crossing the chasm, there is no hope of any external support that is not specifically recruited by you for this purpose.*

## Some Real-World Examples

To see how this works out in actual practice, let's turn now to some specific industry examples. Basically, there are two types of scenarios we want to work through—one where there is installed competition, and the other where there is not. In the former case, it is as if one is trying to invade Normandy from England, and the installed market leader is playing the role of the Nazi forces. In the latter, it is as if one had landed on a new continent and decided to set up shop selling wares to the natives. Neither task is for the faint of heart.

### Weitek and the Abacus Coprocessor

To begin with the competitive example, what more daunting foe to take on in the semiconductor industry than Intel? Yet that is precisely what a small company called Weitek decided to do. The product in this case is a chip for IBM PCs that accelerates their ability to perform math functions on very large or very small numbers—what is called floating point mathematics. The promise of this part, called the Abacus, is that it can make a PC-based workstation system—say, built up around a Compaq Deskpro 386—as effective for computer-aided design as a workstation based around a Sun or Apollo computer, costing perhaps three to five times as much.

The Abacus entered a *math coprocessor* market, however,

which had already been established by Intel with its 287 and 387 products (for the 286 and 386, respectively). All the software that could use a coprocessor on these platforms—essentially, all the math-intensive applications—had already adapted to this architecture. Intel's value proposition, to be sure, was no match for Weitek's in terms of pure speed. But it had a strong whole product advantage. The key to going forward was to find a niche market where a small band of customers would be willing to drive forward an alternative standard and the necessary whole product infrastructure around it.

The selected target market segment was the mechanical engineering community—and, in particular, those engineers interested in three-dimensional renderings. The marketing program began by recruiting partners from the hardware community. Compaq, and following its lead, most other IBM PC–compatible manufacturers, designed into their motherboards a socket for the Weitek chip—even though this was in addition to yet another socket needed for a math accelerator chip from Intel. Now, at least, there was a home for the chip. The question was, What else do we need to deliver the whole product for this value proposition?

The list is not trivial:

1. The CAD software that the mechanical engineer uses—be it Autocad, Versacad, Cadkey, or whatever—had to be modified to support the Weitek chip. This means the software vendors now have to maintain *two versions* of their software—one that runs with the Intel chip, another that runs with the Weitek. In other words, this requires a committed partner. Weitek put its best support engineers on the task. At product launch it had some 16 software vendors committed to the port.

   To put in perspective the depth of this challenge, however, it was some 18 months later that the first major MCAD vendor actually shipped Weitek-supporting product, and vendors are still coming on line some three years later.

2. In order for the software applications vendors to recompile their products to support the Abacus, one or more compiler vendors must first have modified their language compilers to output Abacus-specific object code. This had to

happen before the application vendors could even start; hence, it was an early action item for the product marketing manager.

Weitek recruited first-rate allies here in Metaware and Pharlap. Their C compilers came out on schedule and with high quality. As a result, the Abacus got an unpredicted boost in sales due to technology enthusiasts writing custom C applications of their own.

At the same time, another whole product glitch surfaced. Math-intensive applications deal with massive arrays of numbers. These in turn make heavy demands on memory management. Intel's 386 chip provided a new type of relief from memory constrictions, but to take advantage of it, applications had to use a new "protected mode" provided by the Microsoft DOS operating system. The existing MCAD applications were not designed to take advantage of this mode, and converting to it proved tricky indeed. Because this conversion was critical to the future of the Abacus, Weitek product support engineers ended up becoming experts in this area. This was, in other words, yet another whole product service Weitek took responsibility for.

3. A sophisticated MCAD system needs a number of engineering-specific peripherals, including a high-resolution monitor to represent an adequate level of detail complexity on the screen. The IBM PC's standard for high resolution monitors is VGA—good enough for the office, but not for CAD. So you need a custom monitor and a special video display card to interface between that monitor and the PC proper. In addition, the MCAD workstation, depending on its use, may require four to eight times more memory than its office counterpart, in order to keep calculations on massive matrices of numbers from paging in and out to disk. If the software incorporates work with scanned images or "bit maps," the system will need additional mass storage, either from an optical disk or from a removable system such as Iomega's Bernoulli Box.

All this, it turned out, had no impact on the Abacus directly, but it did create a significant indirect result. Because of the specialized hardware requirements, and the need for some systems integration expertise to pull all the parts of a PC CAD workstation together, it developed that the PC retail channel—even Compaq's, which is as good as it gets—was not up to the challenge of serving the MCAD community. Instead, CAD VARs emerged as the distribution channel of choice. This, in turn, required Weitek to reconceive its whole advertising and marketing communications plan, which had been based originally on using the same distribution channel as supported the Intel 387—namely, a subset of the PC retail channel.

4. As with any new system, people are going to need training. In the case of CAD systems, the trainers need to be familiar not only with the way the software commands work but also with the kinds of designs the end users are creating. Additionally, the system will eventually have to interface with some kind of manufacturing system, so that the designs can automatically be transformed into bills of materials for fabrication. This implies at minimum a file transfer and/or database translation routine. And finally, the system will have to be maintained to keep current with new releases of the CAD software as well as potential upgrades in other components that would lead to greater speed, and hence greater productivity.

All these factors ended up reinforcing the CAD VAR, with its close ties to the software application providers, as the principal ally in developing the Abacus whole product infrastructure.

As this book goes to press, school is not completely out on the Weitek story, but in general, despite all these preparations, the product has not become a mainstream-market success. One reason why this is so has to do with a breakdown in partnering, one that will be described later in this chapter. In addition, however, two new competitors have entered the game, both with Intel-compatible architectures. One of these, a product from a company called ITT, is going after the 387 business with a better price. The other, a product from Cyrix, threatens Weitek's business with speed claims that rival the Abacus.

The interesting question in all this is *why*. The market niche for 3-D rendering is not a big one today—why all the interest?

The answer is that, long term, the ability to manipulate three-dimensional constructs may well underlie the graphical user interfaces that are proliferating across all desktop platforms today. This is by no means a sure thing, but if the market goes in that direction, then owning the de facto standard for high-speed floating point math acceleration means owning a very important mainstream market indeed.

In the meantime, however, Weitek has managed to capitalize on its Abacus efforts in another area—by winning the design-in for math acceleration on the SPARCstation from Sun. Here Weitek has won the role of the incumbent—against another semiconductor giant, Texas Instruments—and is enjoying a mainstream position out of the box. But Weitek could not have achieved this position—it essentially beat T.I. out by achieving a staggeringly fast time to market with the Sun device of less than a year, start to finish—if it had not had the Abacus experience to draw on.

## Document Image Processing

Now let's turn to the other scenario for crossing the chasm, the one where (good news) there is no enemy fortifying the shore against invasion because (bad news) nobody thinks there is anything there to defend. Here the vendor must create a market out of nothing. Here the pragmatist buyers who are the key to the mainstream market do not reject the new product so much as simply watch it for signs of development. They don't say no, in other words; they just don't say yes. Talk about extended sales cycles!

In this situation, entrepreneurs are fighting a race against time. Like the intrepid explorers and colonists of the 16th and 17th centuries, they have landed in terra incognita and have a fixed amount of supplies (working capital) to see them through to self-sufficiency. The question is not whether someday someone will make a successful colony; the question is whether it will be them, or whether they will die in the attempt.

The overall application in the example we will look at is called *document image processing*. The compelling reason to buy behind this application is that, for all the data that businesses and governments have put in computers, they store a hundred times that amount on paper. Paper is omnipresent, and the amount of it in our lives is growing, not decreasing. Despite universal acknowledgement of this fact, despite the fact that the phrase "paperless office" has been with us for the better part of a decade, computer vendors, to date, have in fact done almost nothing to help us handle the problem.

What would help look like? Imagine the following: Letters arriving at the office mailroom are opened there and immediately transformed into electronic form by a scanner. Optical character recognition (OCR) software automatically reads each letter and classifies it in terms of the addressee and the sender; then the letter and time of receipt are registered with the system, and the electronic image is routed automatically to the computer of the recipient. When the person receiving the letter sits down at her computer, she opens her mail with the push of a button and reads it on the screen. Because the resolution of the screen is 300 to 400 dots per inch (the same as a good laser printer), with special low intensity lighting, there is no eye strain from this process.

The addressee, having finished the letter, can file it or trash it or forward it, via electronic mail, to someone else in the office. If the correspondence relates to an ongoing file, she can ensure cross-referencing potential by attaching any number of keywords to the file's header record. Then it goes into optical storage, not into a bulky file cabinet, where it stays until it is needed again.

Outgoing correspondence is treated in a similar way. Rather than send around computer files, which require that there be compatible software at the other end, documents are transformed into the universal "bit map" form that is the basis for any fax or any copier output. They then can be faxed automatically from the author's PC or, if desired, printed on paper and sent via conventional (but increasingly obsolete) means.

There's a lot more to this story—electronic signatures, vast search-and-correlate capabilities, cross-correlation of documents and computer databases on the same PC screen, and so on—but we have enough to at least begin thinking through the whole product. Let us assume, for this exercise, that the target customer is an insurance office. And let us assume that the product to be marketed is a document image processing workstation, of the sort advertised at one time by Wang or made by such small, entrepreneurial companies as Document Technologies Inc. And finally, let us assume that the compelling value proposition for this target market segment is to increase the speed and reduce the cost and error rates of processing transactions involving voluminous correspondence, as in the insurance claims area, for example.

Such applications already exist in some visionary installations. But the mainstream market as yet is undeveloped. As a product manager for a document image processing workstation chartered to cross this chasm, your first task would be to define the whole product. Here are some of the main elements:

1. The first thing needed is a standard "record type" that all parts of the system can recognize and process in an appropriate way. For example, take dots per inch. If the printer wants 300 dpi and the monitor wants 400 dpi and the application software wants 600 dpi, then there will be a lot of time spent thrashing (the literal term, I believe, is *dithering*) back and forth.

   This standard not only must hold across our own internal product line. It should also have broad industry support, so that the mainstream pragmatist buyer can entertain a choice of suppliers and not be locked into a single, proprietary system vendor's standards.

2. The overall system will require a whole new class of management software—document management software—to take responsibility for version control, backup scheduling, security, and confidentiality, not to mention customized workflow routings for each department and application. Without this software, the workstation cannot be integrated into an overall operation productively. Given this new software, the standard needs for training and support also apply.

   No one has ever written such software. In concept, it appears relatively straightforward. But the pragmatist buyers are not interested in concepts—they want the reassurance of referenceable installations. Because if it turns out not to be straightforward, they want somebody else to have spent the money to discover that fact.

3. Document images, even in a compressed state, are considerably larger than normal record types. If document images are to be moved back and forth over a network, the bandwidth and speed of the network need to be substantially increased.

   This has devastating implications for the development of the document image processing market. Networks are just now becoming accepted into the mainstream market—the "Year of the LAN," forecasted for every year throughout the 1980s, finally did arrive late in the decade. Now is not the time to rock the boat with new sets of demands. Now is a time, rather, for rolling out proven systems into department after department, faster and faster, cheaper and cheaper, all to the same standard. Pragmatists are not going to want to attack a new LAN standard for some time to come.

4. Document image systems will need to interact with standard computer records systems, so that transactions in one realm can be reflected in the other. This in turn implies that the workstation will need windowing software, one window into the computer mainframe—emulating a dumb terminal—the other opening up the document image, with commands and utilities to aid in cross-referencing data from the two systems.

   This requirement has significant implications, not only for the amount of software development involved but for the number of interfaces it implies to software standards already in place. And some of these standards are not yet settled. For example, in the Unix environment there are several windowing standards fighting for dominance. If the developers of document image processing want to move faster than the windows standardization movement, then they must support all major contenders. Otherwise they must wait. Neither alternative is attractive.

5. Finally, if we are serious about electronic paper, then its legal status must be confirmed. In particular, the legal validity of an optically stored document and of an electronic signature needs to be established. Otherwise, a paper copy of everything must be maintained.

About this time, as product marketing managers, we are

thinking about circulating our résumés. This is an impossible problem to solve, particularly at the level of an entrepreneurial company seeking to achieve a mainstream market presence. The breadth and scope of the issues that come up when trying to deliver document image processing as a whole product are simply too great. At one end of the spectrum, they are tangible issues—get other pieces of software and hardware to collaborate with our product, creating a product infrastructure that fills out the entire need. At the other end of the spectrum, there is a host of intangibles—standards to agree on, legal issues, policies, and the like. Unless these issues are dealt with authentically, the market simply cannot develop. How can we proceed?

Once again, the initial key is to pursue a highly niched market. At the outset we must find one type of insurance claim processing that we can "own" soup to nuts. Ideally, this type of processing will be confined to a single department, so that we can put it up on our "own" LAN. It will interface to the rest of the world through one gateway, and ideally that will be to a very standard system—either an IBM PC or an IBM mainframe. This specific niche application must support a high ROI for document image processing—as, for example, a Medicare claims processing unit, where handwritten medical records must be collated with computerized hospital care records to produce a claim that can be filed with the U.S. government so the hospital and the doctor can get paid. The longer such filing takes, the more errors in the process, the bigger the "float" of unpaid claims that has to be financed by the hospital and the doctors.

So let us say, at the end of the target market segment selection process described in the previous chapter, we decided to specialize only in Medicare claims. Where does that put us? Well, for one thing, we can now simplify and standardize the workflow software we will support—it will no longer be generic workflow design software but will instead implement a relatively standardized Medicare claims workflow. We can write that software ourselves, or have a Medicare-specialist software consultant develop it. In either case, it can ship with the system, thereby speeding the time it takes for the customers to start realizing the benefits of their purchase. At some later date we might hope to involve the federal government in our standards process, to speed their end of the claims processing—much as the income tax consulting industry is working with the IRS in standardizing submission of computerized tax returns.

Second, since we are only going to work with one type of department, we can deliver our system more or less turnkey, using whatever LAN system and whatever windowing system we choose. True, we will have to support terminal emulation for the industry-leading set of hardware platforms that support hospital billing systems. This means some market research on our end. The good news is, once we find out who they are, we may discover that the leading application vendors want to become partners and allies—our workstation adds value to their product—and the leading hardware vendors who support these applications may later become our first OEM customers for our workstation.

This brings us to the final key to crossing the chasm in a market development scenario where there has been no pioneering work done before us—to work in combination with the right set of partners and allies. In particular, if we, as a small, creative entrepreneurial company, can establish the right relationship with a larger, more established firm, one with the kind of resources and clout that can bull through some of the more tangled issues in whole product development—then we have set up a strong one-two punch.

To bring this example to a close, the first job of the successful product manager is just to draw the simplified whole product diagram—a segmented doughnut, if you will— and label its parts, each segment being a component of the whole product that does not come with the generic product. The guide here is the library of target-customer scenarios, for embedded in each of these characterizations are the implicit assumptions of the whole product. It is simply a matter of thinking through each scenario to visualize the component parts of the overall solution that will be required.

There then follows a make versus buy versus partner decision process for each of the segments in the whole product doughnut. The object is to assign primary responsibility for each segment to someone we can hold accountable—either ourselves, the customers, or a marketing partner. If the choice is the customers, we need to build in the proper level of support for their efforts. If the choice is a marketing partner, we need to go out and recruit. This is what we meant earlier by organizing the marketplace around the whole product. How to go about doing that is our next topic.

## Partners and Allies

Marketing partnerships and strategic alliances are very trendy items in high-tech marketing these days. One expects to see ads in the *Wall Street Journal* any day now reading:

> Large, well-heeled company with established distribution channels and aging product line seeks small, entrepreneurial, cash-starved technology leader with hot new product. Photos available upon request. Write box no. . . .

As as rule, however, these types of alliances do better in the boardroom than on the street. To start with, the company cultures are normally too antithetical to cooperate with each other. Decision cycles are wildly out of sync with each other, leading to enormous frustration among the entrepreneurs and patronizing responses from the established management. To make matters worse, each side has probably misrepresented itself one way or another during negotiations, such that there is plenty of ammunition for each group to fire at the other once tempers get hot. This is particularly likely to be the case when the entrepreneurs have been using acquisition as essentially a financial exit strategy. So, for the most part, despite the impeccable logic of these mergers, they are very tough to bring off.

Of course, some strategic alliances have been extremely successful. Consider the standards movement in the computer industry, where organizations like X/Open and Unix International are drawing together such natural enemies as H-P and DEC, AT&T and IBM, Oracle and Relational Technology, all because the customer cannot achieve whole products without the implementation of industrywide standards. Or consider the relationship between Intel, IBM, and Microsoft that drives the core of the PC industry, or the partnership between Apple and Microsoft in which they actively collaborated to build the Macintosh market even in the midst of a mutual lawsuit.

Powerful as these relationships are, however, the complexities of developing and maintaining such strategic alliances are enough to daunt all but the most megalomaniacal. They are certainly not the province of mere product managers seeking to ensure that their customers achieve their compelling reasons to buy.

What does work for product managers, on the other hand, are tactical alliances. *Tactical alliances have one and only one purpose: to accelerate the formation of whole product infrastructure within a specific target market segment.* The basic commitment is to codevelop a whole product and market it jointly. This benefits the product manager by ensuring customer satisfaction. It benefits the partner by providing expanded distribution into a hitherto untapped source of sales opportunities.

These alliances work because a common interest in the same target customer has drawn together companies naturally. They are already bumping into each other in the same market segment, so they are not strangers to each other when overtures are made. Big Eight consulting firms and IBM mainframe software companies are a good example of such a natural alliance, since the former are often asked to evaluate, select, and sometimes install packages made by the latter. Microcomputer manufacturers like Apple and IBM welcome third parties making add-in boards. Vendors of successful PC software packages, like Lotus 1-2-3, Excel, WordPerfect, and, perhaps the consummate example, HyperCard, help small companies market their third party add-in programs. Why? Because in every case, only by the cooperation of many parties can customers get the whole product they require.

To give one final example, consider the Apple II. In an era when everyone else is complaining about shorter product life cycles, the Apple II outlived its natural life cycle several times over. During some of its best-selling years, it was by any normal measure a thoroughly obsolete product—any measure, that is, but the customers'. Customers simply refused to let the Apple II die. Why? Because it was perhaps the archetypal whole product, having established itself in not one but two sites—the K–12 educational system and the home office. For these sites there still remains a wealth of software, a well-established user-group base, widespread training and support expertise, relatively sophisticated follow-on products built upon a standard platform (Appleworks), a broad range of peripherals, and even, as long as Apple will continue to support it, a relatively sophisticated upgrade path. Apple II users are not a high-tech group. They don't care about being state-of-the-art. They want something that is proven, and once they get it, they won't let go. Surely there is a lesson here for the rest of us.

These types of alliances can be readily initiated and managed at the product marketing manager level. Often, the initial opportunity is first brought to the company's attention either by the salespeople or by customer support staff, one of whom has bumped into the potential ally at a particular customer's site. But they can also be anticipated through the exercise of thinking through the whole product solution to the customer's buying objective. The main point, again, is that these are tactical alliances growing out of whole product needs, not strategic alliances growing out of . . . well, whatever strategic alliances grow out of (my personal feeling is that the number-one cause of strategic alliances is too many staff people with not enough to do).

To see how this works out in practice, let us go back to the case of the Weitek Abacus. The key partners needed to provide the whole product for their MCAD target customer were

1. the PC hardware vendors, led by Compaq, who designed in the socket into which one can plug an Abacus

2. the compiler vendors, principally Metaware and Pharlap, who created the software that allows programs written in C or Fortran to utilize the Abacus as an accelerator

3. the MCAD application vendors, principally Versacad and Cadkey, who ported and recompiled their applications to run in a Weitek-supported mode

4. the MCAD VARs who integrated all of the above into a single working system, which they then sold, serviced, and supported.

All these relationships were set up and managed by the Abacus product manager, who, depending on the size of the partner, was dealing with anyone from a manager of third-party relations to the president. What made all this viable was that the companies already had a common stake in serving the

MCAD customers, specifically assuaging their insatiable appetite for more number-crunching performance. All discussions, therefore, were grounded in the reality of fulfilling a known market need and the benefits of taking a cooperative approach to doing so. Because the scope of reference was legitimately tactical, handling the relationships via the product manager proved appropriate and effective.

Now, having said all that, there was one situation where the wheels did come off, which did escalate to the executive level, and which ultimately created a significant setback for the Abacus product marketing effort. Since we often learn more from mistakes than successes, this makes for an illustrative example.

Early on in the Abacus effort, Autodesk's Autocad was targeted as the number-one application partner, having upwards of 60 percent market share on PC CAD platforms. Unfortunately, the initial approach did not go well. The expert at Autodesk did not believe the performance gain estimates from Weitek, and in the discussions that followed, hot egos, rather than cool heads, prevailed. The problem was escalated to the executive level, as a result of which Compaq was persuaded to intervene on Weitek's behalf at Autodesk. The porting project was then pushed through, despite the objections of the group who, among other things, had to do the port. When the port was completed, the results were disastrous—far from gaining in performance, the software actually ran slower. This, in turn, was followed with considerable finger pointing and I-told-you-so's, with the ultimate effect being a virtual breakdown in communications between the two companies that took more than two years to repair.

In the meantime, the consequences of this split ran far deeper than had originally been imagined. It turned out that not only was Autocad the market share leader in PC CAD sales, it was virtually omnipresent on all PC CAD systems. That is, even when the system was running an Autodesk competitor, it still had Autocad installed as well. This meant that the PC CAD platform had to incorporate both an Intel math chip (to accelerate Autocad's performance) and a Weitek chip (for the other applications). In other words, the price/performance proposition could no longer be based on the difference in price between the two coprocessors, but rather had to be based on their sum.

Moreover, since the socket for the Weitek chip was built as a superset of the existing Intel socket, there was typically only room for one coprocessor, and since Intel's was supported by all vendors, it ended up getting the nod more often than not. Thus, what was originally conceived as a very hot product indeed—the sort of product hit that Weitek subsequently has had with the SPARCstation, turned out to be a more marginal proposition.

The lesson here is twofold: First, the ecology of a whole product is delicate and can be disrupted easily. Second, it is crucial to maintain long-term relationships despite the pressure for short-term results. Weitek and Autodesk could probably have worked something out to their mutual advantage had the relationship stayed open—that is what they are doing today—but two years is a long time not to make progress in a market as dynamic as PC CAD.

To sum up, whole product definition followed by a strong program of tactical alliances to speed the development of the whole product infrastructure is the essence of assembling an invasion force for crossing the chasm. The force itself is a function of actually delivering on the customer's compelling reason to buy in its entirety. That force is still rare in the high-tech marketplace, so rare that, despite the overall high-risk nature of the chasm period, *any company that executes a whole product strategy competently has a high probability of mainstream market success.*

## Recap: Tips on
## Whole Product Management

Again, in keeping with our intent to recap key ideas at the close of each chapter through a tactical checklist, here are eight tips on whole product management:

1. Use the doughnut diagram to define—and then to communicate—the whole product. Shade in all the areas for which you intend your company to take primary responsibility. The remaining areas must be filled by partners or allies.

2. Review the whole product to ensure it has been reduced to

its minimal set. This is the KISS philosophy (Keep It Simple, Stupid). It is hard enough to manage a whole product without burdening it with unnecessary bells and whistles.

3. Review the whole product from each participant's point of view. Make sure each vendor wins, and that no vendor gets an unfair share of the pie. Inequities here, particularly when they favor you, will instantly defeat the whole product effort—companies are naturally suspicious of each other anyway, and given any encouragement, will interpret your entire scheme as a rip-off.

4. Develop the whole product relationships slowly, working from existing instances of cooperation toward a more formalized program. Do not try to institutionalize cooperation in advance of credible examples that everyone can benefit from it—not the least of whom should be the customers.

5. With large partners, try to work from the bottom up; with small ones, from the top down. The goal in either case is to work as close as possible to where decisions that affect the customer actually get made.

6. Once formalized relationships are in place, use them as openings for communication only. Do not count on them to drive cooperation. Partnerships ultimately work only when specific individuals from the different companies involved choose to trust each other.

   In particular, OEM or VAR agreements are only as valuable as the energy the people in the distribution channel are willing to devote to them. Getting and sustaining the attention of someone else's sales force is a full time job, since helping to sell someone else's product is an unnatural act that must be restimulated continually.

7. If you are working with very large partners, focus your energy on establishing relationships at the district office level and watch out for wasting time and effort with large corporate staffs . Conversely, if you are working with small partners, be sensitive to their limited resources and do everything you can to leverage your company to work to their advantage.

8. Finally, do not be surprised to discover that the most difficult partner to manage is your own company. If the partnership really is equitable, you can count on someone inside your company insisting on taking a bigger share of the benefit pie. In fighting back, look to your customers to be your truest and most powerful allies.

# 6

# Define the Battle

On the eve of our invasion, let us regroup. We have already established the point of attack, a target market segment for which we provide a truly compelling reason to buy. We have already mapped out the whole product needed to deliver on this value proposition and have recruited the necessary partners and allies to deliver it. The major obstacle in our way now is competition. To succeed in securing our beachhead we need to understand who or what the competition is, what their current relationship to our target customer consists of, and how we can best position ourselves to force them out of our target market segment.

This is what we mean by defining the battle. *The fundamental rule of engagement is that any force can defeat any other force—if it can define the battle.* If we get to set the turf, if we get to set the competitive criteria for winning, why would we ever lose? The answer, depressingly enough, is because we don't do it right. Sometimes it is because we misunderstand either our own strengths and weaknesses, or those of our competitors. More often, however, it is because we misinterpret what our target customers really want, or we are afraid to step up to the responsibility of making sure they get it.

How far must one go to serve one's customers? Well, in the case of crossing the chasm, one of the key things a pragmatist

136

customer wants to see is strong competition. If you are fresh from developing a new value proposition with visionaries, that competition is not likely to exist—at least not in a form that a pragmatist would appreciate. What you have to do then is create it.

## Creating the Competition

In the progression of the Technology Adoption Life Cycle, the nature of competition changes dramatically. The changes are so radical that, in a very real sense, one can say at more than one point in the cycle that one has no obvious competition. Unfortunately, as the head of European distribution for Adaptec likes to say, "Where there is no competition, there is no market." By way of introduction, therefore, we need to rethink the significance of competition as it relates to crossing the chasm.

In our experience to date with developing an early market, competition has not come from competitive products so much as from alternative modes of operation. Resistance has been a function of inertia growing out of commitment to the status quo, fear of risk, or lack of a compelling reason to buy. Our goal in the early market has been to enlist visionary sponsors to help overcome this resistance. Their competition, in turn, has come from others within their own company, pragmatists who are vying with visionaries for dollars to fund projects. The pragmatists' competitive solution, in general, is to invest dollars to chip away at problems a piece at a time (whereas the visionaries aspire, like Alexander the Great with the Gordian knot, to cut through them with a single mighty—and mighty expensive— stroke). Pragmatists work to educate the company on the risks and costs involved. Visionaries counter with charismatic appeals to taking bold and decisive actions. The competition takes place at the level of corporate agenda, not at the level of competing products.

That's how competitions work in the early market. It is not at all how they work in the mainstream, in part because there are not enough visionaries to go around, in part because visionaries themselves like to play not in the mainstream but rather out in front of it. Now we are in the true domain of the pragmatist. *In the pragmatist's domain, competition is defined by comparative evaluations of products and vendors within a common category.*

These comparative evaluations confer on the buying process an air of rationality that is extremely reassuring to the pragmatist, the sort of thing that manifests itself in evaluation matrices of factors weighted and scored. And the conclusions drawn from these matrices will ultimately shape the dimensions and segmentation of the mainstream market. IBM PCs, it will turn out, are better for finance, and Macintoshes are better for graphics. DEC's VAXs are best for computer-aided manufacturing, Tandem's Non-Stops for automated teller machines, and Sun's SPARCstation for CAD workbenches. Pragmatist buyers do not like to buy until there is both established competition and an established leader, for that is a signal that the market has matured sufficiently to support a reasonable whole product infrastructure around an identified centerpiece.

In sum, the pragmatists are loath to buy until they can compare. *Competition, therefore, becomes a fundamental condition for purchase.* So, coming from the early market, where there are typically no perceived competing products, with the goal of penetrating the mainstream, you often have to go out and *create your competition.*

Creating the competition is the single most important marketing decision made in the battle to enter the mainstream. It begins with locating your product within a buying category that already has some established credibility with the pragmatist buyers. That category should be populated with other reasonable buying choices, ideally ones with which the pragmatists are already familiar. Within this universe, your goal is to position your product as the indisuptably correct buying choice.

The great risk here is to rig the competition, that is, to create a universe that is too self-serving. You can succeed in creating a competitive set that you clearly dominate, but this set, unfortunately, is either not credible or not attractive to the pragmatist buyers. For example, I might claim that I am the greatest high-tech marketing consultant with a Ph.D. in Renaissance English literature. This claim might be credible, but it is not particularly attractive. On the other hand, I might claim that I am the greatest marketing consultant of all time—an attractive claim, perhaps (although it is not obvious to me how one can be a great consultant and egotistical at the same time) but, in any event, not a credible one.

In high-tech marketing, the sins may not be this egregious, but they are very widespread. I am familiar with products that claim to be leaders in such categories as "artificial-intelligence-based DOS shells," "16-bit graphical user interface utilities," and "high density, impact-resistant removable magnetic storage." These "categories" actually had meaning and value during the early market development for these products, because in each case a visionary could translate the technology component into an opportunity to make a strategic breakthrough. They are meaningless, however, to the pragmatist buyers. Such categories neither relate to their concerns nor emerge from the world in which they work. Moreover, these categories appear specifically designed to exclude from the competitive set the very products the pragmatist is most likely to consider as purchase alternatives. As marketing devices for crossing the chasm, therefore, they are useless.

So, how can you avoid selecting a self-servicing or irrelevant competitive set? The key is to focus in on the values and concerns of the pragmatists, not the visionaries. It helps to start with the right conceptual model—in this case, *the Competitive-Positioning Compass*. That model is designed to create a value profile of target customers anywhere in the Technology Adoption Life Cycle, identify what to them would appear to be the most reasonable competitive set, develop comparative rankings within that set on the value attributes with the highest ranking in their profile, and then build our positioning strategy development around those comparative rankings. Here's how it works.

## The Competitive-Positioning Compass

There are four domains of value in high-tech marketing: technology, product, market, and company. As products move through the Technology Adoption Life Cycle, the domain of greatest value to the customer changes. In the early market, where decisions are dominated by technology enthusiasts and visionaries, the key value domains are technology and product.

In the mainstream, where decisions are dominated by pragmatists and conservatives, the key domains are market and company. Crossing the chasm, in this context, represents a transition from product-based to market-based values.

The Competitive-Positioning Compass illustrates these dynamics:

## The Competitive-Positioning Compass

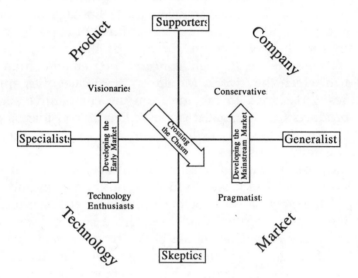

There is a lot of information packed into this model, so let's sort it out piece by piece.

- The directionality provided by the compass comes in the form of the two labeled axes. *The horizontal dimension shows the range of buyer interest in and understanding of high-technology issues.* In general, the early market is dominated by specialists who, by their nature, are more interested in technology and product issues than in market standing or company stature. By contrast, the mainstream is dominated by generalists who are more interested in market leadership and company stability than in the bits and bytes or speeds and feeds of particular products.

- *The vertical dimension overlays a second measure, the buyer's attitude toward the proposed value proposition, ranging from skepticism to support.* Markets begin in a state of skepticism and evolve to a state of support. In the case of the early market, the technology enthusiasts are the skeptical gatekeepers; in the case of the mainstream market, it is the pragmatists. Once they have given their blessings, then their companions—visionaries and conservatives, respectively—feel free to buy in.

- *The model also points to the fact that people who are supportive of your value proposition take an interest in your products and in your company. People who are skeptical of you do not.* This means that, at the beginning of a market, when skepticism is the common state, basing communications on product or company strengths is a mistake. You have no permission to tout these elements, because the market players do not yet believe you are going to be around long enough to make a difference.

- However, there are ways to win over skeptics. *Even the most skeptical specialists are always on the lookout for new technology breakthroughs.* Thus, although you cannot initially get them to sponsor your product, you can get them involved in understanding its technology, and from that understanding, to gain an appreciation for the product itself. The more they appreciate the technology, the easier it becomes for them to support the product.

- *Similarly, skeptical generalists may not take an interest in an unproven company but are always interested in new market developments.* If you can show the generalists that there is an emerging unmet market requirement, one that you have specifically positioned your products and your marketing efforts to meet, then out of their appreciation for the market opportunity, they can learn to appreciate your company.

- *These are the two "natural" marketing rhythms in high tech— developing the early market and developing the mainstream market.* You develop an early market by demonstrating a strong technology advantage and converting it to product credibility, and you develop a mainstream market by demonstrating a market leadership advantage and converting it to company credibility.

- *By contrast, the "chasm transition" represents an unnatural rhythm.* Crossing the chasm requires moving from an environment of support among the visionaries back into one of skepticism among the pragmatists. It means moving from the familiar ground of product-oriented issues to the unfamiliar ground of market-oriented ones, and from the familiar audience of like-minded specialists to the unfamiliar audience of essentially uninterested generalists.

Now let's tie all this back into creating the competition. If we are going to succeed in winning over the lower right quadrant, the skeptical pragmatists, then that competition has to be based in market-oriented concerns. That's what the pragmatists care about. In other words, we must shift our marketing focus from celebrating product-centric value attributes to market-centric ones. Here is a representative list of each:

| Product-Centric | Market-Centric |
|---|---|
| Fastest product | Largest installed base |
| Easiest of use | Most third party supporters |
| Elegant architecture | De facto standard |
| Product price | Cost of ownership |
| Unique functionality | Quality of support |

In the previous chapter, the entire basis of the focus on whole product and partners and allies was to move our leadership premise from the left-hand list to the right. That is, lacking an existing market leadership position, we wanted, within the confines of a manageable market segment, to create the valued attributes of one, and thereby bring a state of true market leadership into existence. Now we need to communicate what we have accomplished so as to win the pragmatist buyers' support.

*To sum up, it is the market-centric value system—supplemented (but not superseded) by the product-centric one—that must be the basis for the value profile of the target customers when crossing the chasm.* This value profile, in turn, will model how the target customers are likely to perceive the competitive set and what position they are likely to accord to a new player coming into that set. Let's see how this works out in a real-world example.

## Creating the Competition:  Novell

It has been a few years since Novell crossed the chasm, thereby legitimizing, after several false starts, the first true "Year of the LAN." At the time it was not obvious that Novell was destined to become the mainstream market leader in LANs. That role looked more likely to go to 3-COM. What happened?

Going into the chasm, 3-COM, with its proprietary hardware platform, had clear advantages over Novell in key product-centric areas, but neither company looked particularly good when measured against a market-centric value profile. There was no industry standard hardware platform. Indeed, there was not even a standard wiring platform underneath for making the physical LAN connections. The traditional leaders in the PC marketplace were being of no help. IBM was not able to provide leadership, despite having been given not one but two chances. Microsoft was finally getting into the fray with LAN Manager but was clearly well behind the power curve. And finally, the distribution channel for anybody's LAN did not appear up to the challenge of providing consistent, no-fault LAN installation and support. No wonder the Year of the LAN was slow in coming.

Within this environment Novell positioned itself primarily against 3-COM, with any other comers welcome, with the following propositions:

1. The Novell solution would be a software-only solution running on top of industry-standard IBM PC architecture.

2. The Novell solution would mask the first two layers of network protocol, thereby giving the customer the freedom to choose from a variety of physical connectivity options.

3. Novell would educate its dealers to a new, Novell-certified standard—the Gold Standard—which would stand for consistent, no-fault LAN installation and support.

There is no question that these decisions entailed further erosion in Novell's product-centric value position. Anytime you support a general purpose hardware platform, rather than one specifically tailored to your own software, there will be a performance cost, and possibly a price disadvantage as well. In this case, however, the mainstream customer was more than willing to pay the performance price just to get reliability. In the mainstream, a system going down is a much greater source of heartburn to pragmatist buyers than one that is a bit sluggish.

On the price front, the Novell solution actually turned up cheaper in a lot of instances, because customers already owned IBM PCs they could free up for network management. Since the hardware platform was already an industry standard, it was easier for Novell to recruit and train dealers who could effectively support their solution. Moreover, the dealer channel did not have to carry an additional load of special inventory, taxing their already overstretched capital resources. And finally, since Novell was not competing for the hardware dollars, it was able to create a powerful de facto alliance with Compaq, a company that had already established a superb position in the mainstream market, and ride their coattails to some degree into becoming the accepted LAN standard.

All these factors became summarized in Novell's fundamental positioning theme of being a software company, unlike 3-COM, or even IBM. That positioning was a function of identifying the pertinent competitive set as 3-COM and IBM, ranking Novell's comparative positioning against them relative to a set of market-centric value attributes, and then taking the high-ground position. *This battle strategy was defined first and foremost by creating the competition in a new context.*

## A Second Example: Quicken

In the case of Novell, there was a very large unmet market need in the LAN arena, one so pressing that there was no longer a need to identify an application niche for getting started. In fact, such niches had long since been developed, and the market was waiting on other issues. In most cases, however, the field is not so open. Let us look at a more common type of chasm crossing, one from a company called Intuit, with a product called Quicken.

Quicken, from a PC specialist's point of view, belongs to a category of software called financial management applications for home use. Today it is the market leader. There was a time, however, when the success of the product—and the company—hung by a thread. How Intuit, and its president Scott Cook, responded to that situation provides a superb lesson in how creating the right competition can accelerate crossing the chasm.

When Quicken was first introduced into the market, the best-selling program was Andrew Tobias's *Managing Your Money*. From a product-centric point of view, it was far richer in functionality than Quicken, offering portfolio analysis and other financial modeling capabilities. To the "financial enthusiasts" who made up the early market for these products, it was clearly the preferred choice, and Intuit was doomed if it continued to play the game on that turf.

In casting about for alternatives, Intuit hit on a very simple value proposition for the home computer user—make it easier to pay bills. This is a pragmatist type of value proposition—we are not making a strategic breakthrough but rather an incremental improvement in a recurrent operation. It is, in other words, a marketing opportunity for the mainstream, not the early market.

Unfortunately for Intuit, there was no established mainstream category called computer-aided bill paying. Pragmatists used checkbooks, thank you very much, and they worked just fine. So how could Intuit penetrate this market?

First, they had to find a manageable market segment. In this case, the world was already pretty well restricted by the qualifier, adults who use computers in their home. The key issue then became checks—if Quicken were to be easier than a manual system, it had to be easy to get the checks. Intuit decided to handle that process for their end user. (As a result, today, income from providing checks is a key component of the company's business strategy, the margins being excellent and the cost of sales virtually nil.) Then a third issue arose—how to align the checks with the printer correctly so that everything printed in the right spot. This turned out to be a major technical problem, ultimately requiring Intuit to invent a subsequently patented solution. Once that was accomplished, however, a true whole product was in place.

Now we come to *creating the competition*. In Intuit's case, it was more a problem of uncreating the competition. The last way it wanted to portray Quicken was as a product for either technology or financial enthusiasts. It wanted to leave the whole realm of "money managers" behind. So it set as its prime competitor *pen and paper*—the manual approach to bill paying.

And it profiled as the key values of its target customer: speed, convenience, and ease of use.

Now taking on pen and paper in the domains of convenience and ease of use is no simple task, and Intuit had to tread very carefully here. The argument, as it finally emerged, however, was compelling. It went something like this. You already use your PC at home. Now, think about how every month you have to write out the same bills to the same people. Wouldn't it be easier to write them out once, and then just update the amount every month, press a button, and have them print out on your home printer automatically? And what about records? Come tax time, wouldn't it be easier if you had all your checks written out of a computerized database, so that you had a complete, computer-sortable record to work with? Might not the product even pay for itself by ensuring you found all the tax deductions you are entitled to? Finally, wouldn't you like to reduce the time you spend paying bills from a couple of hours to a couple of minutes?

The initial answer to all this, in classic pragmatist terms, was maybe. But Intuit was able to secure a beachhead in the home computing market, and from that beachhead to build a word-of-mouth campaign that has ultimately resulted in it being the unquestioned leader in this market. Indeed, in the current wave of IBM PC systems trying to penetrate the home market with bundled solutions, Quicken is appearing as a product of choice to bundle.

*The key point is Intuit's arguments became compelling only after it had redefined the competition.* By creating pen and paper as the competition for itself, Intuit was able to dominate a market segment with a product that was functionally less rich than other products on the market. This strategy was legitimate because Intuit's target customer did not perceive those other products as part of any competitive set it wanted to take an interest in. That is, if you go back to the target customer value profile of *speed, convenience, and ease of use,* a more functionally rich product is normally too inconvenient and too hard to use to capture this type of customer's interest.

So, once again, a unique and successful battle strategy was defined, with its roots beginning in creating the right kind of competition.

# Creating the Competition:
# Some Current Opportunities

So much for looking backward. Hindsight is always 20/20. Let's see what happens when we try to create some competition for products that are just now crossing the chasm. Three worth looking at are PC-based project management software, CD-ROM storage systems, and Prodigy.

## PC-Based Project Management Software

These programs have been around for some time now. Introduced in the wave of software that followed word processing, spreadsheets, and databases, project management was once believed to be the next "killer app." In fact, however, the market never developed beyond the boundaries of the "project management enthusiasts." The leading products today—Symantec's Timeline, Microsoft's Project, Computer Associates' SuperProject, Claris's MacProject, and Software Publishing Corporation's Harvard Project Manager—all vie for a dominant share of a slow-growth segment. They are all in what we might call an atrophied early market—none of them has become a broad mainstream product.

So what would it take? Actually, the situation is very close to the one Intuit was able to solve. The difference is that Intuit was going to have to go out of business if it didn't solve the problem, whereas these companies are all making a modest return, hence are not forced into taking radical creative action. But if they chose to, the most promising path would be to pursue what one might call miniproject management—something a little bit larger than personal time management but not as complex as true project management. The target would be that level of task and resource management that anybody who has a few people reporting to them has to handle. The value profile, in this instance, should be some version of the Quicken proposition: speed, convenience, and ease of use. And like Quicken, there is the added-value proposition that when it comes around to giving annual personnel reviews, there is already a database of record to work from.

Let us assume, then, that one of these vendors decides to

make a run for the mainstream with a new product, called Project Management Lite. Let's assume that the target customer is an established user of a PC or a Mac, and that the product's interface is familiar—either Mac-like or 1-2-3-like. Thus there's not much of a whole product burden on the hardware or the software. There will, however, be a significant education burden: What is miniproject management? How do I do it? And why should I care? So we are either going to handle this ourselves or recruit a partner or ally to do it.

Let us suppose all this is in place. What kind of competition do we want to create? If we apply the value profile to the target customer, the competitive set of alternatives for miniproject management software is likely to be the following:

1. I do it in my head.
2. I do it with pen and paper.
3. I do it with a spreadsheet and electronic mail.

People committed to (1) and (2) are going to be too hard to win over, and probably do not have a sufficiently complex burden to warrant a software solution. So let's try to position ourselves against the spreadsheet and e-mail combination, which will be our entrenched mainstream competitor.

The competition has a number of established advantages we would be foolish to challenge. Their software is already in place and, after a fashion, it works. The spreadsheet, once you label all the columns by time intervals, makes a serviceable Gantt chart for task deadline management, and the e-mail is keeping everyone in touch. Where is a gap we can exploit?

One gap is the inability to "roll up" miniprojects across groups to create a larger view of the organization's functioning. Another is the lack of integration between this form of project management and, say, a time-billing system. A third is the lack of any resource management dimension at all, meaning that a person could go unutilized without it being drawn to the attention of the manager in charge. The net result of all of these gaps, particularly in an organization that bills out its professional services, is "leaking revenue" or "creeping unbillability," either one of which one could claim to fix with the right kind of miniproject management software.

So, let's redefine our target segment as professional service firms with project work that is not particularly complex to model. Let's redefine the whole product as a project management and billing system. This is going to put new pressure on the whole product—specifically, an interface to the company's accounting software—but let's assume that is manageable. Now, our created competition is "business as usual," consisting of some nonstandard amalgam of computer-aided tools and paper-based notes. Our fundamental value proposition will have very little to do with the esoteric details of project management and everything to do with standardizing the fundamentals of day-to-day operations so that key profitability factors are more readily managed. This is, needless to say, a very different market from the one that is currently defined by project management software, involving a very different battle strategy, based on a very different competitive set.

Okay. Now let's stop. If we look back over the previous few paragraphs spent creating competition and defining a battle strategy, we see that right in the middle we changed our target segment. That, in turn, forced us to rethink our whole product. *In other words, each of the four component elements of the D-day effort—targeting the point of attack, assembling the invasion force, defining the battle strategy, and, as we shall see in the next chapter, launching the invasion itself—interacts with every other.* A change in any one of these four will ripple through the other three. The process for building a marketing plan to cross the chasm, therefore, consists of continuously working through the changes in this set of component elements until one comes up with a worthy opportunity.

## CD-ROM Storage Devices

Let's look at a second example, one that has not been in the marketplace all that long—CD-ROM storage devices. This is the same fundamental technology that those of us who have music-loving teenagers have learned is going to replace cassette tapes—at about twice the price. Now we are discovering it is also available to support computers. It is being positioned with multimedia, where it is a must-have component needed to handle the very large data objects that are associated with reproducing realistic visual images. As such it is an important player in a developing early market.

At the time this manuscript is being drafted, it is probably still too early for CD-ROM vendors to be thinking seriously about crossing the chasm—the early market itself is still under development—but let us suppose that we set the clock ahead a year or two and give ourselves the problem of turning this very exciting technology into a mainstream computer product.

I can envision at least four wildly differing scenarios, each of which is driven by what competition the CD-ROM vendor decides to create. I offer each of these with a "prediction" designed to show how different the results could turn out.

1. One could target magnetic disk drives (called Winchesters, after the original codename for their development at IBM) as the competition and go after such target applications as customer support or catalog sales. The boundaries of the market would always be defined by a price/performance gap on large bit-mapped objects. Optical technology has a substantial advantage in cost per byte, which increases as the number of bytes increases. On the other hand, providers of Winchester technology have a whole product infrastructure already in place, have consistently been able to achieve substantial cost reductions, and show no immediate signs of "hitting the wall," so they represent a truly tough competitive set, particularly if the applications are as cost-sensitive as the ones just suggested.

   *Prediction:* Following this competitive set will lead to insufficient market success, ultimately causing the enterprise to go out of existence.

2. One could target minicomputer and mainframe systems devoted to engineering and manufacturing documentation management. The current trend toward downsizing across all platforms would work in favor of anyone selecting this competitive set. The downside is that optical storage is also a relatively easy extension to make to a large scale system environment, and minicomputer and mainframe vendors field strong account management teams who are good at fending off competitive forays onto their turf.

   *Prediction:* Following this competitive set will lead to permanent niching in a few, probably highly regulated markets, where documentation and change order management are critical success factors. Elsewhere the mainstream competitor will keep the CD-ROM vendor from ever securing a beachhead.

3. One could target as the competitive set the world of microfiche—specifically, the world of distributed, updated reference documents, catalogs, and databases. The idea here would be to leverage computer-aided search and retrieval software, to facilitate broad or complex data searches, and to target applications where supplementing the output with images provides greater added value. The downside here is the current lack of whole product infrastructure, the amount of capital investment required to move from no computerization to a CD-ROM-level capability, and the cost sensitivity of many of these applications. Additionally, users of microfiche tend to come out of the precomputer office automation days, tend to be conservatives in relation to computer usage, and therefore are not likely to buy into revolutionary technology change.

*Prediction:* This is a real mainstream market opportunity for a very low-cost, dedicated CD-ROM reader system combined with an end-user programmable CD-ROM create-and-update workstation. But because this depends on significantly lowered manufacturing costs due to prolonged high-volume experience in making the devices, this will come only well after CD-ROM has achieved mainstream market success somewhere else.

4. One could target the competitive set of VCRs and videotapes, both in the areas of entertainment and education. This would be a high-end niche market for starters, based on interactivity with an image-based curriculum or entertainment. The biggest challenge would be securing a sufficient volume and quality of titles to drive the adoption of enough machines to lower the cost of production so one could gain enough market share to get more titles, and so on. The downside is that stimulating interactivity may not be as compelling as mind-numbing passivity—a discouraging but amply supported claim.

*Prediction:* This is most likely to be driven by the entertain-ment industry as the form factors for music, music videos, and motion picture videos converge into one. For the com-puter industry, it will probably be a cul-de-sac.

Now, the purpose of these predictions is not to enter my hat into the small ring of luminaries who really do look out into the future of the computer industry. Rather, it is to create a form to express pragmatic hypotheses for judging the value of market-ing alternatives. And there are two points to be made in so doing. First, *predictions themselves are meaningless unless they are understood in the context of a particular target market and a particu-lar competitive set.* All this is just another way of saying that cre-ating the competition at the time of crossing the chasm is fre-quently the single most significant marketing decision made in the life of a product.

The second point: *There is no winning scenario predicted.* This is the most common outcome of marketing brainstorming ses-sions. It should not be viewed with either surprise or alarm. It is the norm. Winning scenarios are hard to find, and for even the most promising candidates, there are far more reasons one can assemble for why they won't work than for why they will. Sooner or later you have to pick one or another and try to act it out, but it is very rare that the odds are stacked in your favor. And all this is just another way of saying that high-tech market-ing is a high-risk endeavor, with higher-than-normal rewards balanced by longer than normal odds.

## Prodigy

There is a third type of lesson to be learned about creating com-petition, and the current marketing of Prodigy exemplifies it perfectly. The product, cosponsored by IBM and Sears, is soft-ware for the IBM PC which, when used with a modem and a phone, hooks the end user into a nationwide computer net-work. You might think of this network as akin to a TV net-work—it has a variety of programming and services to choose from, from getting Dow Jones news feeds to making travel arrangements to doing home shopping to consulting on-line encyclopedias to getting 900-number information about things like weather or sports. It is designed as a product for home use and, among other things, has been bundled into IBM's PS/1 computer for the home market.

At the time of this writing, Prodigy is actively seeking to establish itself in the mainstream market. In addition to bundling itself in with the PS/1, it is currently running TV ads with the tag line, "You just gotta have this thing!" Based on everything that has been said thus far in this book, this is a *wrong strategy*. Here's why.

*First, there is no competitive set*. Pragmatists, we said, won't buy until there is a clear competitive context. What is Prodigy competing against? In any one of its ads, you get a shotgun of minicompetitors—the *Wall Street Journal*, travel agents, the *Encyclopaedia Britannica*, TV news, 900 numbers, etc. This is not a competitive set. That is, you do not buy those five things together as one thing. Hence, there is nothing here for Prodigy to compete against directly.

*The lack of a direct competitive set also makes it difficult to articulate a compelling reason to buy*. Again, pragmatists find their compelling reasons to buy in incremental improvements to important operations. They see the new product providing a superior substitution for some old way of doing business. It is not clear how any one of these services is focused on offering such an improvement.

*Further, the ads have little appeal for pragmatists*. Rather, their appeal is much more in line with the values of visionaries. They are trying to create a vision of a new kind of world, much in the same sense as the IBM PS/1 ads running at the same time, where you ride a zoom camera through every room in the house and see how every member of the family is benefiting from the new computer. The problem here is that TV advertising is neither an efficient nor an effective way to reach visionaries. Remember, they are looking for the order-of-magnitude improvement effect. This is not the sort of thing that can be convincingly communicated in a 30-second spot. Also remember, they are going to take their technological direction from the technology enthusiasts, and that group tends to reject all consumer-oriented advertising as marketing hype.

So, for a variety of reasons, Prodigy appears to have been marketed in the wrong way. It is reasonable for the company to be targeting entry into the pragmatist mainstream—both the product and the whole concept of on-line services have had sufficient exposure to warrant the move. *The fastest way out of the thicket they're in is to recreate their competition.*

Let's see if we can't help by eliminating some contenders. First off, drop travel agents. They provide a whole raft of added value services beyond mere reservation making, at no extra charge, and few pragmatists in their right mind are going to want to forego those services. And second, drop home shopping. Don't drop the service—for that is paying a lot of the bills—but don't put home shopping in the primary competitive set, the one around which we will position the product. Why? Shopping is entertainment for most people, an opportunity to fantasize and to be seduced. The computer is not a very rich medium for supporting either the fantasy or the seduction, not as effective as the home shopping TV programs, for example, where you can see the products live and call in and talk with someone who will, well, seduce you.

What does this leave? It leaves a variety of information services. Now, in general, these services have to date been marketed by "information enthusiasts," that is, people who think that more information is better than less. There is considerable evidence, however, that the opposite is true. Pragmatists, in particular, have highly developed strategies for screening out information rather than letting themselves get swamped by it. At the same time, pragmatists appreciate the value of rifle-shot selection of a key piece of data at the time when they want it.

If we turn to the pragmatists at home, what kind of information applications might they want that they can't easily get some other way? The encyclopedia is a good one. They are expensive and they go out of date. If the on-line access were essentially free—which Prodigy, by its fixed-price approach to subscription, makes possible—then the notion of an adult, or better a child, browsing an electronic encyclopedia is great, particularly if there are pictures that come through with adequate resolution (which, unfortunately, just added an expensive chunk to our whole product, but that's another story). A comprehensive library of cookbooks and recipes might be great. Cousin Harry has just dropped off two pheasant—how do you cook pheasant? And this would lead to a bulletin board for cooks, led or monitored by someone like Narsai David or Julia Child, where people could interactively exchange ideas and tips. (This bulletin board communication has been the most widely accepted aspect of Prodigy to date, leading among other things, to the company having to step in and edit some communications—a disastrous marketing move—but that is also another story.)

Yellow pages. White pages. The Dow Jones database for business research. Math textbooks and tutoring. English. Chemistry. The list is endless. The competitive set becomes the current set of information services available to the American family, with special emphasis on the American child. The positioning would be based on the inability of the current set to support the child's education adequately, with the added value of great support for the parents if they choose to use it.

Now, all of this has a huge impact on the whole product. Bringing any serious database on line is a gargantuan undertaking made even more complex by having to present the information in a way that a novice can navigate it. Hence the need to *target the point of attack*. Even Sears and IBM are not big enough to enter the mainstream in a broad swipe. But if they picked, say, freshman through junior year of high school, and focused on the top five textbooks nationwide in each of the fundamental course areas, that becomes a targetable point of attack. The compelling reason to buy is to make an incremental improvement in a child's education and ability to go on to college. The key partners and allies become the textbook manufacturers, the PTAs and school boards, and the teachers themselves.

As an ex-educator, I find this positioning particularly attractive, but we could target other points of attack around other segments. What we cannot do is bludgeon our way into the mainstream without some kind of focus.

## Positioning

Creating the competition, more than anything else, represents a watershed moment in positioning. Positioning is the most discussed and least well understood component of high-tech marketing. You can keep yourself from making most positioning gaffes if you will simply remember the following principles:

1. *Positioning, first and foremost, is a noun, not a verb.* That is, it is best understood as an attribute associated with a company or a product, and not as the marketing contortions that people go through to set up that association.

2. *Positioning is the single largest influence on the buying decision.* It serves as a kind of buyers' shorthand, shaping not only their final choice but even the way they evaluate alternatives leading up to that choice. In other words, evaluations are often simply rationalizations of preestablished positioning.

3. *Positioning exists in people's heads, not in your words.* If you want to talk intelligently about positioning, you must frame a position in words that are likely to actually exist in other people's heads, and not in words that come straight out of hot advertising copy.

4. *People are highly conservative about entertaining changes in positioning.* This is just another way of saying that people do not like you messing with the stuff that is inside their heads. In general, the most effective positioning strategies are the ones that demand the least amount of change.

Given all of the above, it is then possible to talk about *positioning* as a verb—a set of activities designed to bring about *positioning* as a noun. Here there is one fundamental key to success: When most people think of positioning in this way, they are thinking about how to make their products *easier to sell*. But the correct goal is to make them *easier to buy*.

Companies focus on making products easier to sell because that is what they are worried about—selling. They load their marketing communications with every possible selling argument, following the age-old axiom that if you throw a lot of mud at a wall, some of it is bound to stick. Prospective customers shrink from this barrage, which in turn causes the salespeople to chase after them that much harder. Even though the words appear to address the customers' values and needs, the communication is really focused on the seller's attempt to manipulate them, a fact that is transparently obvious to the potential consumer. It's a complete turn-off—all because the company was trying to make its product easy to sell instead of easy to buy.

Think about it. Most people resist selling but enjoy buying. By focusing on making a product easy to buy, you are focusing on what the customers really want. In turn, they will sense this and reward you with their purchases. Thus, easy to buy becomes easy to sell. The goal of positioning, therefore, is to create a space inside the target customer's head called "best buy for this type of situation" and to attain sole, undisputed occupancy of that space. Only then, when the green light is on, and there is no remaining competing alternative, is a product easy to buy.

Now, the nature of that best-buy space is a function of who is the target customer. Indeed, this space builds and expands cumulatively as the product passes through the Technology Adoption Life Cycle. The are four fundamental stages in this process, corresponding to the four primary psychographic types, as follows:

1. *Name it and frame it.* Potential customers cannot buy what they cannot name, nor can they seek out the product unless they know what category to look under. *This is the minimum amount of positioning needed to make the product easy to buy for a technology enthusiast.*

   Discontinuous innovations are often difficult to name and frame. The largely ineffectual category called *groupware* is an attempt to name and frame a new class of personal computer software products designed to enhance the productivity of groups. As a category, however, its boundaries are so ill-defined, and its essence so elusive, that it might better be called *soupware*. This, in turn, has had a negative impact on any product that has been positioned in this category. *Names* and *frames* must be real and reasonable to work.

2. *Who for and what for.* Customers will not buy something until they know who is going to use it and for what purpose. *This is the minimum extension to positioning needed to make the product easy to buy for the visionary.*

   Object-oriented databases, pen-based laptops, ISDN networking, the Intel i860 microprocessor, and Lotus's Notes are all in the process of seeking out this type of positioning in order to build up their *early market.* As it becomes clear who can most benefit from these technologies to achieve a major strategic advantage, then they will have secured the necessary positions to develop their respective early markets.

3. *Competition and differentiation.* Customers cannot know what to expect or what to pay for a product until they can place it in some sort of comparative context. *This is the minimum extension to positioning needed to make a product easy to buy for a pragmatist.*

Examples of this category have filled the preceding pages of this chapter. The key is to provide the reassurance of a competitive set, and of a market-leading choice within that set.

4. *Financials and Futures.* Customers cannot be completely secure in buying a product until they know it comes from a vendor with staying power who will continue to invest in this product category. *This is the final extension positioning needed to make a product easy to buy for a conservative.*

Apple, for example, has only recently passed into this realm. This, combined with the release of their new low-end Macintosh Classic, gives Apple for the first time a compelling proposition for targeting customers in the back half of the Technology Adoption Life Cycle. IBM, on the other hand, has long dominated this domain.

The purpose of positioning is to put in place these sets of perceptions with the appropriate target customers in the appropriate sequence and at the appropriate time in the development of a product's market.

## The Positioning Process

When positioning is thought of primarily as a verb, it refers to a communications process made up of four key components:

1. *The claim.* The key here is to reduce the fundamental position statement—a claim of undisputable market leadership within a given target market segment—to a two-sentence format outlined later in this chapter.

2. *The evidence.* The claim to undisputed leadership is meaningless if it can, in fact, be disputed. The key here is to develop sufficient evidence as to make any such disputation unreasonable.

3. *Communications.* Armed with claim and evidence, the goal here is to identify and address the right audiences in the right sequence with the right versions of the message.

4. *Feedback and adjustment.* Just as football coaches have to make half-time adjustments to their game plans, so do marketers, once the positioning has been exposed to the competition. Competitors can be expected to poke holes in the initial effort, and these need to be patched up or otherwise responded to.

This last component makes positioning a dynamic process rather than a one-time event. As such, it means marketers revisit the same audiences many times over during the life of a product. Establishing relationships of trust, therefore, rather than wowing them on a one-time basis, is key to any ongoing success.

Positioning is at the heart of everything we do at Regis McKenna Inc., and, as a process, it is well documented in a book called *The Regis Touch* (Addison-Wesley, 1985). It is not my intent to go over that same ground here. There are specific aspects of positioning, however, that do merit some extra attention when looked at in the specific context of crossing the chasm.

## The Claim: Passing the Elevator Test

Of the four components, by far the hardest to get right is the claim. It is not that we lack for ideas, usually, but rather that we cannot express them in any reasonable span of time. Hence the elevator test: Can you explain your product in the time it takes to ride up in an elevator? Venture capitalists use this all the time as a test of investment potential. If you cannot pass the test, they don't invest. Here's why.

1. *Whatever your claim is, it cannot be transmitted by word of mouth.* In this medium the unit of thought is at most a sentence or two. Beyond that, people cannot hold it in their

heads. Since we have already established that word of mouth is fundamental to success in high-tech marketing, you must lose.

2. *Your marketing communications will be all over the map.* Every time someone writes a brochure, a presentation, or an ad, they will pick up the claim from some different corner and come up with yet another version of the positioning. Regardless of how good this version is, it will not reinforce the previous versions, and the marketplace will not get comfortable that it knows your position. A product with an uncertain position is very difficult to buy.

3. *Your R&D will be all over the map.* Again, since there are so many different dimensions to your positioning, engineering and product marketing can pick any number of different routes forward that may or may not add up to a real market advantage. You will have no clear winning proposition but many strong losing ones.

4. *You won't be able to recruit partners and allies*, because they won't be sure enough about your goals to make any meaningful commitments. What they will say instead, both to each other and to the rest of the industry, is, "Great technology—too bad they can't market."

5. *You are not likely to get financing from anybody with experience.* As just noted, most savvy investors know that if you can't pass the elevator test, among other things, you do not have a clear—that is, investable—marketing strategy.

So how can we guarantee passing the elevator test? The key is to define your position based on the target segment you intend to dominate and the value proposition you intend to dominate it with. Within this context, you then set forth your competition and the unique differentiation that belongs to you and that you expect to drive the buying decision your way.

Here is a proven formula for getting all this down into two short sentences. Try it out on your own company and one of its key products. Just fill in the blanks:

- *For* (target customer)
- *Who* (statement of the need or opportunity)
- *The* (product name) *is a* (product category)
- *That* (statement of key benefit—that is, compelling reason to buy)
- *Unlike* (primary competitive alternative)
- *Our product* (statement of primary differentiation).

Let's try a few examples. Suppose we are Amdahl, a maker of plug-compatible clones of IBM mainframe, and let us say that our primary competitive alternative is Hitachi Data Systems. Our elevator message might be:

> For Fortune 500 companies who are looking to cut costs and who operate in data centers of IBM mainframe computers, Amdahl's computers are plug-compatible mainframes that match or surpass the equivalent IBM computers in features and performance, at a far more attractive price. Unlike the Hitachi line of computers, our products have been backed by the same service and support organization for over 20 years.

Now, what is often interesting about writing a statement like this is not what you write down but what you have to give up. In particular, when it came to competitive differentiation, I picked stability of the service and support organization as opposed to any number of other possible differentiators. Wouldn't it be better to load in several different ones for a bigger effect?

The answer here is an emphatic *no!* Indeed, this is just what defeats most positioning efforts. *Remember, the goal of positioning is to create and occupy a space inside the target customers' head*. Now, as we already noted, people are very conservative about what they let you do inside their head. One of the things they do not like is for you to take up too much space. This means they will use a kind of shorthand reference: Mercedes ("top-of-the-line, conservative"), BMW ("upscale performance sedan, yuppie"), Cadillac ("American top-of-the-line, tired") Lexus ("New kid on the block, current best buy"). That's all the space you get for your primary differentiation statement. It's like a telegram with less than one line. If you don't make the choice to fill the space with a single attribute, then the market will do it for you. And since the market includes your competition trying to unposition you, don't count on it to be kind.

Let's try another example, this one from the software side:

For IBM PC users who want the advantages of a Macintosh-style graphical user interface, Microsoft Windows 3.0 is an industry-standard operating environment that provides the ease of use and consistency of a Mac on a PC-compatible platform. Unlike other attempts to implement this type of interface, Windows 3.0 is now or will very shortly be supported by every major PC application software package.

Note how much more powerful this positioning is than that for Windows 2.2 of the previous year, a product that was still vying with OS/2 Presentation Manager for dominance of the desktop. At that time, the second sentence in the statement would have gone something like this:

Unlike OS/2 Presentation Manager, Windows is less resource-demanding and is available today.

Arguments like this are necessary when one is skating past a particularly difficult competitor, but they are hardly compelling.

One final point on claims before moving on to other issues: *The statement of position is not the tag line for the ad.* Ad agencies come up with tag lines, not marketing groups. The function of the statement of position is to control the ad campaign, to ensure that however "creative" it may become, it stays on strategy. If the point of the ad is not identical with the point of the claim, then it is the ad, not the claim, that must be changed—regardless of how great the ad is.

## The Shifting Burden of Proof

The toughest thing about high-tech marketing is that just about the time you get the hang of something, it becomes obsolete. This is even true of something as innocent as providing evidence. That is, like everything else in high tech, the kind of evidence that is needed evolves over the course of the Technology Adoption Life Cycle. This can be summarized within the structure of the Competitive-Positioning Compass:

## Positioning: The Evidence

By working your way up the left and then up the right of the compass, you can trace the evolution of desired evidence as the market evolves from the technology enthusiast to the visionary to the pragmatist and conservative. The key point to notice is the transition from product to market, corresponding to crossing the chasm. This is simply a corroboration of a point we have been making all along, that pragmatists are more interested in the market's response to a product than in the product itself.

What is particularly awkward for a high-tech company making this transition is that for the first time the major sources of desired evidence are not directly under its control. This is not a matter of having the right features or winning the right benchmark war. It is a matter of other people—theoretically disinterested third parties—voting to endorse your product through not only words but deeds. It is actual investment in building the whole product that demonstrates to the pragmatist that if you are not already the market segment leader, you are destined to become so.

*In sum, to the pragmatist buyer, the most powerful evidence of leadership and likelihood of competitive victory is the quality and number of partners and allies you have assembled in your camp, and their degree of demonstrable commitment to your cause.* The kind of evidence this buyer is looking for is signs of comarketing, such as joint sales calls and cross-referencing each other's products in sales literature, and consistent mutual support even when the other party is not present in the room.

This point leads directly into communications strategy for crossing the chasm. Not only do you have to develop this kind of evidence of whole product support; you also have to make sure that everyone hears about it.

## Whole Product Launches

The concept of a *whole product launch* is a derivative of the widely known practice of a product launch. That is, whenever a new high-tech product is introduced, it is customary to launch it by first briefing the industry analysts and long-lead press editors well in advance of the launch date (so they can serve as references), and then taking the top company executives on a tour to the weekly trade press the week prior to announcement, with the announcement itself capped by an event.

These product launches work just fine when the product itself is "new news." Then, they are an appropriate tool for the development of early markets. By the same token, however, they are not appropriate for crossing the chasm. At this point the product is not new news—at least it had better not be if we are planning to win over the pragmatist buyer. The trade press is not interested, therefore, in a great trumpeting article on Release 3.0 (not unless, that is, you are a Microsoft, but that's another story). So if the message is not "Look at my hot new product," then what is it, and how are you going to get it out?

The message now is "Look at this hot new market." The message typically consists of a description of the emerging new market, fed by an emerging set of partners and allies, each supplying a part of the whole product puzzle, to the satisfaction of an increasingly visible and growing set of customers. The lure embedded in this story is that we are seeing a new trend in the making, and everyone who has a seat on this bandwagon is going to be in on The Big Win. This is a great story for small entrepreneurial companies to be able to tell, because it gives them a credibility that they cannot achieve on their own. Their product does not even have to be at the center of the puzzle—it just has to be an indispensable piece, like Microsoft's operating system, or Conner's or Seagate's hard disk drives, or the Lotus 1-2-3 spreadsheet. The idea is, once you're in, you're in.

Now, how can marketing communications improve your odds? First, marketers have to pick the right communications venue. There are two venues, in general, that lend themselves to whole product stories. The first is the business press. Whole product stories, particularly ones sparked by partnerships and alliances coming together to bring off some wonderful result for a particular company, are the bread and butter of business fare. Companies organizing to bring off this feat consistently, and thereby dominate a particular market segment, are particularly of interest. If the company is brand new, to be sure, the business press is leery. In this instance it is important first to build some references in the financial analyst community, based not on the company per se but on the market opportunity it has in its sights. Financial analysts are usually quite open to briefings on emerging market opportunities, and in that context, can be wooed to take an interest in an emerging entrepreneurial venture. Once they have bought into the market, then they can be used as a reference point by the business press in developing a story.

In bringing this story to the business press, it is important to bring along as many of the other players in the market as possible. One effective tactic is to hold a press conference with multiple spokespersons on the dais—customers, analysts, partners, distributors, and so on. A more elaborate version of the same approach is to sponsor a conference on the core issue that is driving the development of this market. The key objective in either case is to communicate the bandwagon effect in progress.

Finally, communicating via the business press has to be done within the framework of a big idea. Technology stories, told at the level of technology, are only interesting as vignettes, squibs for the column that leads the second section of the *Wall Street Journal*. For a technology story to be a *business story*, it has to be about something that transcends high tech. Typically, the seed of the story is either a new type of opportunity or problem that can now be addressed effectively because of advances in the industry. These advances will have been sparked by technology breakthroughs, and that will be part of the story, but they are now seen to extend to the entire whole product infrastructure, and that will be the main thrust of the story.

The great benefit of the business press as a medium of communication is its high degree of credibility across virtually all business buying situations. This is a two-edged sword for the entrepreneurial company. In order to preserve its credibility, the business press is reluctant to endorse entrepreneurial enterprises until they have been well proved. It takes a long time, in other words, to earn coverage. On the other hand, having broken through in this medium once, it is much easier to do so again. Furthermore, subsequent product-oriented coverage in the trade press tends to become more thorough as the company attains greater stature in the business press.

So building relationships with business press editors, initially around a whole product story, is a key tactic in crossing the chasm. In addition to the business press, the other communications channel for getting out a whole product message is what could be loosely termed "vertical media"—that is, media specifically dedicated to a particular industry or a particular profession. Industry trade shows and conferences, meetings of professional associations, and publications dedicated to a specific market segment all tend to attract pragmatists and conservatives, people who put a high value on maintaining relationships within their group. These associations are relatively open to participation from supporting vendors, provided that the vendors are not too obtrusive with their sales messages.

Whole product issues are ideal for this kind of communications. The idea is to get in a room with a number of people in a given industry and outline the current state of affairs in the vendor's marketplace as it relates to their business. Correctly framed, these sessions put the customer, rather than the vendor or the vendor's product, at the center of things. They align themselves with the customer's needs and the alternatives available to meet those needs. Thus, although they are at one level clearly self-serving to the vendor, they do not *feel* self-serving, positioning the vendor more as a consultant than as a salesperson.

The goal of a whole product launch campaign, overall, is to

develop relationships in support of a positive word-of-mouth campaign for your company and products. The first thing to remember is that developing these relationships takes time— time to ferret out who are the key influencers, time to get to know them on more or less equal footing, time to get up to speed on the industry issues so that the relationship is pertinent and valuable to both parties. The other thing to remember is that, once these relationships are in place, they represent a major barrier to entry for any competitor. Pragmatists and con- servatives—the core of any mainstream market—like to do business with people they know.

## Recap: The Competitive-Positioning Checklist

To define the battle effectively so that you win the business of a pragmatist buyer, you must:

1. Focus the competition within the market segment estab- lished by your must-have value proposition—that is, that combination of target customer, product offering, and compelling reason to buy that establishes your primary reason for being.

2. Create the competition around what, for a pragmatist buyer, represents a reasonable and reasonably comprehen- sive set of alternative ways of achieving this value propo- sition. Do not tamper with this set by artificially excluding a reasonable competitor—nothing is more likely to alienate your pragmatist buyer.

3. Focus your communications by reducing your fundamental competitive claim to a two-sentence formula and then man- aging every piece of company communication to ensure that it always stays within the bounds set out by that for- mula. In particular, always be sure to reinforce the second sentence of this claim, the one that identifies your primary competition and how you are differentiated from it.

4. Demonstrate the validity of your competitive claim through the quality of your whole product solution and the quality of your partners and allies, so that the pragma- tist buyer will conclude you are, or must shortly become, the indisputable leader of this competitive set.

# 7

# Launch the Invasion

In this chapter the final pieces of the D-day strategy come into play—distribution and pricing. As we launch our invasion across the chasm, distribution is the vehicle that will carry us on our mission, and pricing is its fuel. These two issues are the only two points where marketing decisions come into direct contact with the new mainstream customer. Decisions in both distribution and pricing, therefore, have enormous strategic impact, and, with distribution in particular, there is typically only one chance to get it right. For this reason, we have put these two last in our invasion planning sequence, so that we could have the advantage of nailing everything else down first.

*The number-one corporate objective, when crossing the chasm, is to secure a channel into the mainstream market with which the pragmatist customer will be comfortable.* This objective comes before revenues, before profits, before press, even before customer satisfaction. All these other factors can be fixed later—but only if the channel is established. Or, to put it the other way around, if the channel is not established, nothing further can be accomplished. Finally, given that establishing the channel is the number-one goal, the fundamental function of pricing during this same period is to achieve this same end. In other words, during the chasm period, the number-one concern of pricing is not to satisfy the customer or to satisfy the investors, but to *motivate the channel*.

To sum up, when crossing the chasm, we are looking to attract *customer-oriented distribution*, and one of our primary lures will be *distribution-oriented pricing*. These are somewhat radical words, and they are going to require considerable discussion. In order for this discussion to make any sense, however, we need first to review the somewhat troubled state of distribution in high tech as we enter the last decade of this century. There is, at present, a structural problem in the distribution function for high tech, one that may well work itself out eventually, but probably not in time to solve any of our immediate problems. Once we understand that problem, then we can better chart our chasm-crossing strategy.

## The Structure of High-Tech Distribution

There are currently a wide variety of distribution channels operating under the umbrella of the high-tech market. The most prominent are the following:

- *Direct sales.* Typically national in scope and focused on calling on major accounts, this consists of a dedicated sales force in the direct employ of the vendor, with no other intermediary between the customer and the company. IBM has the most famous direct sales force in the world.
- *Retail sales.* This has typically been a two-tier distribution channel, although current restructuring is causing a lot of changes here, with large distributors taking product from vendors and reselling it to retail outlets, who in turn have the direct customer contact. Businessland, Computerland, and Egghead are well-recognized names in this category.
- *Industrial distributors.* This is a one-tier channel, used heavily by the semiconductor industry for standard parts, to provide component parts to computer and computer-related equipment suppliers. Companies like Arrow and Anthem fit in this category.

- *VARs (value-added resellers).* This is typically a two-tier channel, wherein large distributors pass product on to operations that integrate a variety of component products, often including their own proprietary software, into a complete, and typically application-specific, product for the customer. These are typically "no-name" companies, the most successful of which average around 15 employees and $15 million a year in sales revenues.

- *OEMs (Original Equipment Manufacturers).* This is at least a two-tier transaction, beginning with a direct sales force selling to manufacturers, who then integrate the purchased product into their own systems, and sell the systems on to the customer. If the OEM product is bought through industrial distributors and sold through retail or VARs, there can actually be as many as four tiers to this channel. The big computer manufacturers, just like the big auto makers, all purchase products on an OEM basis from the rest of the industry.

- *Systems integrators.* This is not a channel per se, since it rarely if ever sells the same products twice. Rather it is a project-oriented institution for managing very large or very complex computer projects. Since, however, such projects often "design in" standards that are then replicated throughout the rest of the company's operations, there are good reasons to treat systems integrators like a channel.

In recent times we have seen new variations on these basic themes, such as VADs (value-added dealers—essentially retail storefronts providing VAR-like systems integration functions, within a very restricted scope of choices, for the retail customer), super-VARs (wherein local VAR businesses have been purchased and assembled into a nationally organized network, such as EverNet), mail order (a very low cost retail-like outlet, optimized for selling commodity PC products), and superstores (very large retail outlets like Comp USA and Fry's—the Toys "Я" Us of the computer industry—where the primary added value is a broad selection of brand choices at low prices).

Now, from the point of view of crossing the chasm, all these channels can be sorted out by how they stack up against the following three factors:

1. *Demand creators versus demand fulfillers.* Direct sales forces, for example, are optimized for creating demand, while retail superstores are optimized for fulfilling it. A lot of other channels commit to do both, are optimized for neither, and suffer the consequences.

   When crossing the chasm, our immediate goal is to create mainstream demand, but we must also look ahead toward putting in place a channel that can fulfill it.

2. *Role in providing the whole product.* Systems integrators and VARs are optimized for playing a very large role in providing or developing the whole product, and make much of their profits from this service. By contrast, retail and OEM channels take a low-cost position, based on the assumption that the whole product is already "institutionalized" and can be fully assembled from off-the-shelf parts. Again, there are a number of channels "caught in the middle," the most visible of which, at the time of this writing, are high service retail outlets like Businessland.

   In the chasm case, the goal is to take the burden of whole product off of the channel in order to free it up to spend more time creating—and fulfilling—demand for the product.

3. *Potential for high volume.* In some ways, this is simply the obverse of the previous category. Channels optimized for whole product development are not effective for high volume delivery. There is too much labor in their business mix, so that when business is booming, they tend to slow down their selling efforts to work off some of the backlog—thereby flattening what could otherwise be meteoric growth. The low-cost, low-service channels are just the opposite. Optimized for high volumes, they are great for boom times, but they do not do well in start-up mode and tend to panic and dump when business softens, thereby trashing not only their own margins but yours as well.

   In the chasm case, our ultimate target is likely to be a high-volume channel. This is something like getting a car into high gear. The question is, How do we get it up to speed?

What we are looking for, in general, as we cross the chasm, is the following: First and foremost, does the channel already have, or is it optimized to create, a relationship with our target mainstream customer? If not, then it is not a candidate for help-ing us cross the chasm. If, nonetheless, it is our customer's ulti-mate preferred form of distribution, then we are going to have to look for a two-step process, where we have an intermediate distribution tactic to create the relationship and a longer-term one to reap maximum rewards.

Second, how will this channel fit into our whole product mix—our partners and allies strategy? The less pressure we put on the channel to deliver the whole product, the more it can focus on selling instead of supporting. On the other hand, it is absolutely critical that our mainstream customer gets the whole product, and we should be willing to sacrifice some volume in order to prevent customer dissatisfaction at getting less than the whole product.

With these factors in mind, let's take a closer look at some of the more prominent channels in operation today, and specifical-ly, how they stack up against our immediate goal of crossing the chasm.

## Direct Sales

Historically, the most consistently successful channel in high tech has been the direct sales force. More than anything else, it was IBM's mastery of this medium that drove it to prominence, and then dominance, in the 1960s and 1970s. Other companies have successfully copied this act: Computer Associates and Cullinet in mainframe systems software, Oracle and Ask in soft-ware for midrange database applications, Sun in workstations, and Hewlett-Packard, Tandem, and DEC for minicomputer hardware.

The direct sales force is optimized for creating demand. At its center is a consultative salesperson who works with the client in needs analysis and then, supported by a team of application and technology specialists, develops and  proposes solutions, which, after additional interaction with the customer, and a competitive procurement, turn into purchase orders. This is a very expensive way to sell, with the cost of sales built into the product's price. It works reasonably well when two conditions are met.

For the customer, the key condition is that the vendor supply a broadly comprehensive and reasonably competitive set of offerings. If this condition is not met, it means that additional vendor interactions are needed. There is only so much time and effort the customer is willing to put into educating and negotiating with vendors, so breadth of product line is crucial here.

For the vendor, the key condition is both the volume and the predictability of revenues. To support a single consultative salesperson requires a revenue stream of anywhere from $500,000 to several million dollars, depending on the amount of presales and postsales support provided. Say our quota is $1,200,000. That means we must close $100,000 per month. If the sales cycle is six to nine months, and if we are able to close one out of every two opportunities, then we have to have either 12 to 18 $100,000 prospects in the pipeline at all times (not very probable) or some smaller number of significantly larger deals going.

One underlying point here is that there is a price point below which this method of distribution cannot work. If we have to be working $500,000 opportunities, we cannot be selling a product whose base price is $20,000. It turns out the practical limit for base-product price point is around $50,000, with variations depending on the level of selling support required and the speed and predictability of the sales cycle.

Another underlying point is the importance of what salespeople call *account control* but which might more accurately be termed *account cooperation*. Direct sales forces can bring lots of service to an account, but not if they lose the deal during the competitive procurement. Basically then, for this system to work, there has to be a fundamentally *uncompetitive* agenda operating, a you-scratch-my-back-and-I'll-scratch-yours agreement under which vendors are granted a limited monopoly, subject to their not exploiting it egregiously and continuing to provide premium service. This confers a high degree of predictability of revenue and a lower cost of sales.

*When functioning at its best, within the limits just laid out, direct sales is the optimal channel for high tech. It is also the best channel for crossing the chasm.* Nonetheless, it is currently under heavy fire from a number of different directions.

First, wherever vendors have been able to achieve lock-in

with customers through proprietary technology, there has been the temptation to exploit the relationship through unfairly expensive maintenance agreements topped by charging for some new releases as if they were new products. This was one of the main forces behind the open systems rebellion that undermined so many vendors' account control—which, in turn, decreases predictability of revenues, putting the system further in jeopardy.

A second consequence of the open systems competition has been the dramatic increase in the relative value or "price/performance" of computers, and the consequent drop in average selling price. This is further exacerbated by the "commodity status" open systems confers on a computer, driving out the opportunity to preserve margins through proprietary differentiation. As price points lower, it becomes increasingly difficult to sell through a direct sales force. This, in turn, puts heavy pressure on companies with direct sales forces already in place—the minicomputer companies in the Northeast are a prime example—to field sufficient product to generate the kind of revenue volumes needed to maintain this high overhead channel.

Working against the success of that effort is the fact that the complexity of total solutions has increased to the point where no single vendor can cover a big piece of the pie. This undermines the primary customer benefit of account cooperation with a direct sales force—the simplifying of vendor relationships and the improved accountability of working with a single vendor. Now there are too many cooks involved, and major accounts are looking to other kinds of channels—notably, systems integrators—for this kind of overall problem simplification.

Now, against this background, let us look at direct sales as a distribution alternative for crossing the chasm. To qualify at all, our product must have an appropriate pricepoint, so let's assume it does. We like a direct sales approach because it is optimized for creating demand, something very much on our minds. What then are the issues to consider?

Our first issue is, Can we get our sales force entry into the pragmatist's restricted domain? Obviously, if you hire good enough people and hammer loud enough at the door, you can gain some level of entry. What we really mean here is, Do we have a partner or ally who already has a relationship with our target customer and who can help open the door from the inside? Without that sort of leverage, a direct sales force, particularly in its first year of existence, can be very expensive indeed, as you can end up paying very high wages to people who are essentially spending most of their time doing low-grade prospecting—and resenting every minute of it.

An alternative to consider here, where there is an "inside" partner, is recruiting the partner to become our distribution channel. The good news is that the relationship of trust is already in place. The bad news is that it is very hard to capture and maintain the involvement of someone else's sales force. Nonetheless, it can be done, and this approach will be discussed later in this chapter under "Selling Partnerships."

Our second issue is, Do we have the capability to recruit and grow a direct sales force appropriate to the market opportunity? It is certainly possible for a company to create outstanding early market success and not have any significant kind of sales force management capability. Indeed, such capabilities may be countercultural to the firm. If such is the case, then setting out to build a direct sales force can be a very dangerous proposition. An all too typical failed scenario begins with bringing in one or more high-priced, highly ambitious sales executives, who, in trying to create a winning sales situation, run roughshod over the existing culture, politicize the management environment, create divisiveness both within the executive staff and between that group and its investors, and, in general, reduce the effectiveness of the team just at the time when it is being most challenged.

A reasonable alternative here is to field a direct sales force as a transition-oriented tactic, with a long-term goal to take the product into a different channel, through a selling partnership. This will represent a major reduction in overall return on investment—since he who owns the customer owns the profit margins and the future of the product—but it also represents a major reduction in risk, and potential grief. This is not the macho high-tech way, of course, which in my view gives it added attractiveness.

*All other things being equal, however, direct sales is the preferred alternative because it gives us maximum control over our own destiny*. And as we also noted, even if we cannot pass these tests, because creating demand is so important when crossing the chasm, we may well want to copy many techniques from direct sales to supplement, or transition to, whatever distribution channel we finally select.

## Retail Sales

The second most successful channel in high tech is a relatively new entrant on the scene, retail sales, brought into existence by the personal computer and its ever-broadening, highly institutionalized whole product entourage. That is, what the PC brought to the industry for the first time was an open platform with standard hardware and software interfaces. This meant that thousands of vendors could create and supply the parts of an industry-standard whole product, thereby institutionalizing that whole product and opening up the opportunity for retail.

Since its initial appearance in stores like The Byte Shop, retail sales has evolved into a somewhat bewildering variety of forms, including both one-tier channels like Comp USA and Fry's and two-tier systems, the first tier being a distributor like Ingram Micro D or Merisel, the second tier being smaller chains and individually owned and operated outlets. Running in parallel, with a more technical product line, are the traditional distributors of semiconductor components, such as Arrow, with some overlap. Adding to the mix are enterprises with outbound sales forces, such as Computerland and Businessland, who act a lot like direct sales forces, and those with a special service orientation, such as Corporate Software with its focus on helping Fortune 500 clients manage the sheer logistics of distribution and upgrade of PC software products such as the current transition to Windows 3.0. If we are going to succeed in crossing the chasm, we need to step back from this somewhat staggering array of alternatives and look at first principles.

*First and foremost, the retail system works optimally when its job is to fulfill demand rather than to create it.* Unlike direct sales, it does not support the consultative sale. It cannot explain complex software or facilitate complex integration of products. It is not, in other words, well set up to be a participant in developing the whole product. Rather, it is structured to leverage an institutionalized whole product by supporting convenient access to a broad selection of brand choices, providing these choices at the lowest possible prices, and, in the process of so doing, serving as a credit broker among the intermediary parties in the distribution chain.

Now, in a sense, as far as crossing the chasm is concerned, we need go no further. *Because it does not create demand, and because it does not help develop whole products, retail distribution is structurally unsuited to solving the chasm problem.* The overwhelming bulk of retail revenues goes to companies that have been in place since the early 1980s or to companies cloning these leaders' products. The top-selling products in any given year are, for the most part, upgrades of the top-selling products from last year and the year before and the year before that. Of the top 100 software companies in 1989, for example, only 14 had been founded since 1983. Occasionally someone in the channel, shamed by this track record, commits to discover and promote newer products—Egghead loudly touted just such a program in 1989 and then canceled it in 1990. These efforts are doomed, because they require the channel to spend a disproportionate, and ultimately unproductive, amount of time on something that gives too low a rate of return.

The ultimate consequence of this for the retail channel is extremely serious. In 1990 supermarkets introduced 12,500 new products. That is part of retail's added value in the consumer packaged goods market. High-tech retail cannot introduce a tenth that many—because every product carries with it so many whole product demands. Until the industry matures further, retail distribution is in a structural quandary, with far too many players fighting over far too few successful whole products.

So let us rephrase our interest in the retail channel. Let's say we have a product that—*once it is established in the mainstream market*—will be a natural candidate for retail distribution. Now, how should we proceed?

Simply put, we need some intermediary step during which we can create the demand and institutionalize the whole product, and then turn it over to the channel. Some proven approaches include the following:

1. *Direct response advertising.* This works particularly well for

developing demand for low-priced software products, such as Intuit's Quicken, the check-writing package for use in the home or small business, where the risk of trial is not great and the whole product infrastructure is already in place. By varying the pitch in the ads, marketers can sort out which reasons to buy are truly compelling. Once the demand itself is demonstrated, and the pitch proven, then the product can be readily absorbed by the channel.

2. *Telesales (and teleservice).* This works better for higher priced products, like Dell's line of PC-compatible computers, where the company was able to target a particular kind of pragmatist customer—power users—and provide them with better-than-retail service through highly trained and motivated people working over the phone. The low cost of sales was passed on to these customers as lower prices, and because the customers were unusually knowledgeable, even complex product discussions could be accomplished without face-to-face meetings and demos. Again, now that demand for the Dell brand is well established, it is branching out from this system into traditional retail outlets, including Comp USA.

3. *Value-added resellers.* Later we are going to discuss this channel separately, analyzing its suitability for long-term participation in the mainstream market, but here we are looking at it simply as a transitional vehicle. It is an excellent vehicle for developing whole product support, although it is not particularly motivated to package or institutionalize these solutions. It is only fair at creating demand, for, although it uses a consultative approach, it tends to be dominated by problem solvers rather than salespeople, and so sometimes lacks basic selling skills. Nonetheless, for products in PC CAD or higher-end desktop publishing, this is a good channel to get them across the chasm, from which marketers can then migrate them into the retail channel, once demand is established and the whole product in place. Autodesk's AutoCad achieved prominence in this way.

All three of these techniques, then, serve the chasm-crossing strategy by bridging between an immediate need to create demand and/or institutionalize the whole product, after which the product can be turned over to the retail channel in order to leverage that channel's high-volume capabilities.

Now, because this product will ultimately end up in retail, there is an upper limit to its price point. In 1991, this appears to be $10,000. Beyond that point, whole product demands are likely to exceed the retail channel's capacity to supply. Over the coming decade, however, we can expect that pricepoint to go up, *if there are significant advancements in whole product institutionalization.* The automotive industry currently sells $50,000-and-up products through an essentially retail channel, but this is possible only because the whole product for that industry has become so thoroughly institutionalized. In the meantime, for the computer industry, $10,000 is a reasonable cap for retail.

This leads us to an interesting price domain—what do you do between the $50,000 floor of direct sales and the $10,000 ceiling of retail?

## VAR-Land or No-Man's-Land?

Today, the domain between $10,000 and $50,000 is where the structural problem in high-tech distribution is taking its greatest toll. Products in this range provide all the challenges of high-priced products and all the margins of low-priced ones. This creates a very painful squeeze on the vendor, and this squeeze is currently driving a brutal shakeout in the computer industry.

The most visible product line caught in this squeeze is the Unix-based departmental system designed for the commercial market. The old-line minicomputer vendors, forced from their proprietary domains, are trying to distribute these through their direct sales forces. Since the product, however, is approaching commodity status, there is no way to maintain the pricing margins needed to sustain this channel. At the same time, PC vendors migrating up to the Intel 486 chip are manufacturing the same product. Their problem is that there is no way the retail channel can provide the whole product support needed to sustain customer satisfaction with this product. Unlike the PC market, the Unix market does not support a single, industry-standard set of hardware and software interfaces. Until it does, there can be no institutionalized whole product, and therefore no effective use of retail distribution.

One of the most dramatic consequences of this structural gap is that, although every information systems analyst in the world agrees that "client/server" architecture is the only way to go in the 1990s, the client/server market is, in fact, flat on its back—I would argue for lack of an efficient distribution channel. This is not to say that there is no channel for products in the $10,000 to $50,000 price point. The channel is there—it is the VAR channel. It has the expertise to solve the whole product problem. And it has the low overhead to live with tight pricing margins. Indeed, it is a perfect fit. So what's the problem?

The first problem is that *there are not enough VARs to go around.* Because they are relatively small, no-name firms, with relatively high turnover in any given year, not all VARs are listed, and not all listed VARs are in business. Best guess is that there are between 2,000 and 5,000 VARs operating in the United States at any given time, depending how you define the term. The current average annual revenue of the top 250 of these VARs—the ones consistently listed in all the databases—is $15 million per year, or a total annual revenue of about $3.75 billion. The statistical distribution, however, is highly skewed by some very large VARs. By the end of the 250 sample, revenues are averaging only several hundred thousand dollars per year. After this point, good statistics are not currently available, but even if we give the channel the benefit of the doubt in every case, it is hard to build a case for much more than $5 billion annual revenues.

Now, of that $5 billion, how much represents sell through of vendor products versus labor charges that go directly to the VAR? Only the former count in evaluating the VAR channel's effectiveness as a channel. Most VAR managers I have spoken with recently have an approach-avoidance problem with selling hardware. They want to do so because it gets the revenue number up higher, but they also want to get out of it, because the margins are not good. If only half of the volume is labor, that means that the sell through of this channel is under $3 billion. In an industry that is approaching $1 trillion in worldwide annual revenues, this is not a lot of bandwidth. All of which goes to support my first point—VARs are a scarce resource, and good VARs an even scarcer one.

A second problem with the VAR channel is that, *because its best margins come from labor, not product, it tends to sell enough to fill its plate and then stops selling until it gets hungry again.* That is, like any labor-intensive business, once backlog gets to a certain level, the management of the firm tends to focus on working off the backlog rather than getting more new business. This is not the way either a direct sales force or a retail operation works. In both of those cases, the more you sell, the more you want to sell. From a product supplier's point of view, in other words, VAR distribution is an inherently inefficient mechanism, one that resists its own momentum.

There is a corollary to this principle. VARs tend to be people who perceive themselves not as salespeople but as problem solvers. Often technical in orientation, they perceive selling as a necessary evil, what you have to do in order to get the "real work." This service-oriented rather than sales-oriented self-perception results in a channel that is not very good at selling, further contributing to its inefficiency.

For all these reasons, *VARs do not typically make for a good mainstream distribution channel.* Even the exceptions to this rule—Ask Computer Systems, for example, the leader in supplying MRP II systems and also a DEC and H-P VAR, or Novell, the leader in LANs, who until recently also supplied hardware platforms—are making themselves over into software-only companies, as the available margins they can charge for hardware continue to deteriorate. This is not to say that VARs do not belong in the marketing mix—they have high added value in the early market, where developing the whole product is the highest priority. But it does mean that, except for transition tactics, they are not appropriate as the prime channel in a mainstream market success.

## Adaptations and Alternatives

In this category go the remaining distribution alternatives, which include systems integrators, super-VARs, affiliates, OEMs, selling partnerships, outbound retail, and VADs. From a chasm perspective, each is either inappropriate or too specialized to warrant a lot of attention.

### Systems Integrators

These companies have had a long history in the federal government market, where the customer needed all of the advantages of a direct sales relationship, but could not promise a de facto uncompetitive procurement relationship. This meant there was no one to hold accountable for overall systems success. Into this gap entered companies like Electronic Data Systems and Computer Science Corporation.

As the concept of mission-critical systems crossed over from the world of NASA to the world of Fortune 500 boardrooms, commercial America began to take on projects that posed a similar class of problems. At this point, what used to be called the Big Eight firms began to get involved, led by Arthur Andersen. In general, they focus on servicing early market opportunities sponsored by visionary customers, a venue in which systems integrators shine. That is, they run in advance of the institutionalization of the whole product and promise to confer strategic advantage on those customers hardy enough to brave the new technologies. Because they do not serve pragmatist customers, on the other hand, they are not suitable as a prime channel for crossing the chasm.

Systems integrators are, however, an important part of a mainstream marketing program. This is because the design decisions made in the superprojects can set the procurement agenda for years to come. If the Bank of America settles on Tandem for its ATM platform in the pilot project, you can count on a lot of Tandems being sold over the next 5 to 10 years. *Because such design-ins can accelerate mainstream market acceptance dramatically, it is critical for companies crossing the chasm to work in cooperation with systems integrators.*

Unfortunately, most marketing organizations lump this assignment in with VAR and OEM sales, and assign a quota to the whole mix, thereby creating an absolutely indigestible clump of work. As was said earlier, systems integrators are not a channel. They do not sell the same thing twice. They are better viewed as an agent of the customer, and they should be served by a direct sales effort on a project-specific basis by senior executives in the vendor firm. They are not suitable for servicing through normal direct sales, because the length of the sales cycle is typically too long, the probability of winning the deal too low, the likelihood of being kept in the deal after it has been won too uncertain, the ultimate payoff too far out in the future, and the special considerations asked inappropriate for anyone but a senior executive in the firm to grant.

It turns out that the most important marketing contribution to ensuring effective working relationships with systems integrators is a communications task, not a selling one. Commercial systems integrators are typically organized in a partnership structure for selling and doing the business, supported by a centralized advanced technology center. The marketing organization's role should be to keep the centralized organization up to speed on any advanced products it is bringing to market, and to develop partner-level communications access for early warning of emerging customer opportunities.

### Super-VARs

Super-VARs are a very new phenomenon emerging in the LAN market, but one potentially of interest for any product line in the $10,000 to $50,000 no-man's-land. Essentially, super-VARs are an attempt to overlay a nationwide orientation toward finance, technological expertise, and demand creation over a locally oriented demand fulfillment and service channel. In 1990 several venture-funded companies, including EverNet, emerged more or less at the same time with this business plan.

The target customers for this channel are the pragmatist buyers in medium to large-scale companies. These buyers want the security of working with a well-funded organization. Further, they may well have nationwide service issues, so that the LAN they build for the first installation in Pennsylvania may need to be replicated in Alabama and Wisconsin. Most VARs simply cannot operate on a nationwide service basis. Further, on occasion pragmatist buyers will want access to advanced technological expertise. No single VAR can expect to cover the wide breadth of technologies at stake, but a network of many VARS, each contributing its own distinctive expertise, could. Finally, as we noted, local VARs are not as a rule very good marketing organizations. But a national company could invest in marketing expertise, take the demand creation "selling" off the VAR's plate, and leave that organization with the demand fulfillment and service roles for which it is so well suited.

Overall then, the idea is to address some of the inherent limitations in the VAR channel. *On paper the super-VAR channel looks like it should work. School is still out as to whether it makes out in the marketplace.* If it does, this could become an extremely important new channel, especially for the trying-to-emerge client/server market.

### Affiliates

Affiliates are another way to try to skin the same cat. SSA, the number-one supplier of MRP II systems on IBM's midrange platforms (System 36s, System 38s, and AS/400s), built itself into the market leader using this technique.

In an affiliate distribution scheme, the primary vendor, in this case SSA, supplies the base product, along with the national marketing campaign, marketing materials, coselling support, and the like. The affiliate helps to sell and then customizes the base product for each specific customer installation. In the case of the MRP II market, where the product cannot be used until it has been integrated into customer-specific operations, such on-site service is critical. At the same time, no particular local company could be expected to put in the R&D effort necessary not only to build the base product initially but also to keep the product current with advances in technology and methodology.

What is particularly of note in this strategy is the way in which it supports distribution-oriented pricing, about which we say more later. But the main principle is to transfer the margin leverage from the supplier vendor to the distribution channel, thereby winning an unusual degree of loyalty and support from the channel. This can allow companies like SSA to generate mainstream market penetration in record time. *Because it lends itself so well to leveraging distribution-oriented pricing, the affiliate channel is a great approach to crossing the chasm.* It remains to be seen if this channel can transition to mainstream stability.

## OEMs

The logic of an OEM channel is particularly attractive for a small company seeking to do business with hard-nosed pragmatist customers. Why not leverage the direct sales force of an established player in the market?

The difficulty here is in winning the attention of the OEM with a product that requires some creative selling. Demand creation requires sales force focus. The OEM sales force, however, is likely to be focused on the big-ticket products that come out of the company's own R&D labs, not the add-on product coming in from another vendor. Only when that add-on product is sufficiently in demand that its inclusion becomes a deal winner—or the lack of it a deal breaker—will the channel work on the supplier's behalf. That is never going to be the case for a chasm product. *Because it has no patience for the special demands of a chasm product, therefore, the OEM channel is not suited to solving the chasm distribution problem.*

## Selling Partnerships

*Selling partnerships* is not really an industry term, but they are a key tactic in crossing the chasm. The idea is to take the attractiveness of the OEM channel—the notion of leveraging an established relationship with a pragmatist mainstream customer—while recognizing the limited attention one can command in someone else's sales force.

The basic tactic is to cosell with a whole product partner, fielding a direct sales force yourself, sharing leads with the partner, each bringing the other in to help develop comprehensive whole product proposals. In its most noncommittal form, this is not a particularly powerful relationship, but it can be highly leveraged if the chasm-crossing partner devotes resources to evangelizing and educating the other partner's sales force. The key is to simplify the selling arguments so that the partner's sales force has one or two key points to make, generating little subsequent debate from the customer, enough to create an entry point, but not enough to overburden or risk the sale. The right kind of simplification is not always obvious, and it tends to evolve over the life of the product, so success in this tactic requires an ongoing commitment of marketing resources.

Overall, selling partnerships are a transitional strategy that must ultimately resolve into one or another of the stable mainstream-channel alternatives. The reason for this has to do with price. The established partner in this relationship is not willing to introduce another player who has a product that will soak up a lot of the dollars in the procurement. Basically, if the other product costs more than 15 to 20 percent of the established product, there is too great a perceived risk that the price of the total solution will get out of control. That puts some kind of relative cap on the price of product to which this tactic can be applied. At the same time, this is a very high-cost method of selling, consisting essentially not only of a direct sales force but also the marketing support needed for an indirect one. This puts more pressure on price margins, given the relative price cap, than a company can sustain over time. *Selling partnerships, in other words, are good for priming the pump, but not for the long term.*

### Outbound Retail

How do you provide the benefits of a retail channel to someone who does not want to go to the store? Well, if you are Domino's Pizza, you build a franchise based on home delivery. That, in essence, is what outbound retail is all about.

Pragmatist customers, particularly at a Fortune 500 account, want to buy certain types of high-tech products in volume from direct sales representatives who call on them—at least initially—and then go onto deal with the purchasing department. For these purchase decisions, they do not need a consultative sales process, however, nor do they want to pay the heavy margins associated with this type of channel. Outbound sales forces from retail outlets meet these criteria.

*Outbound retail sales forces, however, do not meet the chasm criteria.* Although they are organized and managed like a direct sales force, they are not consultative, and therefore they are not demand creators. They are demand fulfillers. That is the way their compensation works, and that is all that the price margins on their products can sustain. A chasm product is a nuisance to them. It generates a disproportionately large amount of expla-

nation to generate a disproportionately small amount of revenue.

## VADs

The category of value-added dealers has grown up out of many dealers being unable to add value. Frankly, a dealer that cannot add value should be—and very shortly will be—out of business. Therefore, VADs are a noncategory. This is an example not of any restructuring but of vocabulary inflation and hype in a never-ending quest to justify one's share of the margin pie. *Therefore, you can ignore this channel entirely.*

# So What's the Right Choice?

This is the sort of question that, as any good consultant knows, you are not supposed to answer but instead duck by saying, "It depends." Having said this, however, let's see if there is anything more useful to say.

In general, we have had positive ratings for direct sales and affiliates as channels of choice, and negative ratings for retail, VADs, VARs, OEMs, and systems integrators, with the jury still out on super-VARs. These ratings can be offset, of course, by other programs in the marketing mix. For example, retail might work just fine if you had a truly compelling demand creation program that created pull through the retail outlets. Usually, however, you are better off taking one of the transitional alternatives discussed until the demand creation process is well understood and well under way. Conversely, selecting the "right" channel is not the whole answer. Even a very fine direct sales force, or the most inspired set of affiliates, unsupported by a credible market positioning, cannot succeed in crossing the chasm to the pragmatist customer.

Finally, however, there is one last element in the marketing mix that has a major impact on the outcome of chasm crossing. And that is the topic of our next, and final, section in this chapter.

# Distribution-Oriented Pricing

Pricing decisions are among the hardest for management groups to reach consensus on. The problem is that there are so many perspectives competing for the controlling influence. In this section we are going to sort out some of those perspectives and set out some rational guidelines for pricing during the chasm period.

## Customer-Oriented Pricing

The first perspective to set on pricing is the customers', and, as we noted in the section on discovering the chasm, that varies dramatically with their psychographics. Visionaries—the customers dominating the early market's development—are relatively price-insensitive. Seeking a strategic leap forward, with an order-of-magnitude return on investment, they are convinced that any immediate costs are insignificant when compared with the end result. Indeed, they want to make sure there is, if anything, *extra money* in the price, because they know they are going to need special service, and they want their vendors to have the money to provide it. There is even a kind of prestige in buying the high-priced alternative. All this is pure *value-based pricing*. Because of the high value placed on the end result, the product price has a high umbrella under which it can unfold itself.

At the other end of the market are the conservatives. They want low pricing. They have waited a long time before buying the product—long enough for complete institutionalization of the whole product, and long enough for prices to have dropped to only a small margin above cost. This is their reward for buying late. They don't get competitive advantage, but they do keep their out-of-pocket costs way down. This is *cost-based pricing*, something which will eventually emerge in any mainstream market, once all the other margin-justifying elements have been exhausted.

Between these two types lie the pragmatists—our target customers for the chasm-crossing effort. Pragmatists, as we have said repeatedly, want to back the market leader. They have learned that by so doing they can keep their whole product costs—the costs not only of purchase but of ownership as well—to their lowest, and still get some competitive leverage from the investment. They expect to pay a premium price for the market leader relative to the competition, perhaps as high as 30 percent. This is *competition-based pricing*. Even though the market leaders are getting a premium, their allowed price is still a function of comparison with the other players in the market. And if they are not the market leader, they will have to apply the reverse of this rule and discount accordingly.

From the customer perspective, then, as we argued in the previous chapter, the key issue is market leadership versus a viable competitive set, and the key pricing strategy is premium margin above a norm set by comparison.

## Vendor-Oriented Pricing

Vendor-oriented pricing is a function of internal issues, beginning with cost of goods, and extending to cost of sales, cost of overhead, cost of capital, promised rate of return, and any number of other factors. These factors are critical to being able to manage an enterprise profitably on an ongoing basis. None of these, however, has any immediate meaning in the marketplace. They take on meaning only as they impact other market-visible issues.

For example, vendor-oriented pricing typically sets the distribution channel decision by establishing a price-point ballpark that puts the product in the direct sales, retail, or VAR camp. Moreover, once the product is in the market, vendor-oriented factors can make a big impact if, for example, they allow us a low-cost pricing advantage in a late mainstream market, or if they allow us to use operating margins to fund new R&D for the next early market.

Vendor-oriented pricing, however, represents the worst basis for pricing decisions during the chasm period. This is a time when we must be almost entirely externally focused—both on the new demands of the mainstream customer and the new relationship we are trying to build with a mainstream channel. Indeed, because of the primary importance of securing ongoing means of access to the mainstream, this latter issue should be the number-one factor for pricing decisions during this period.

### Distribution-Oriented Pricing

From a distribution perspective, there are two pricing issues that have significant impact on channel motivation:

- Is it priced to sell?
- Is it worthwhile to sell?

Being priced to sell means that price does not become a major issue during the sales cycle. Companies crossing the chasm, coming from success in the early market with visionary customers, typically have their products priced too high. Price does become an issue with the pragmatist customer, but when the channel feeds back prospect resistance and uses comparable products as evidence of the expected pricing, companies too often argue that they have no such competition, and that the channel does not know how to sell the product properly.

However, products can also be priced too low to cross the chasm. The problem here is that the price does not incorporate a sufficient margin to reward the channel for its extra effort in introducing this novelty into their already established relationship with the mainstream customer. If the channel is going to go out of its way to take on something new, the reward has to be significantly more attractive than whatever is available from business as usual.

If we put all these perspectives together and look at them in a crossing-the-chasm context, the fundamental pricing goal should be as follows: *Set pricing at the market leader price point, thereby reinforcing your claims to market leadership (or at least not undercutting them), and build a disproportionately high reward for the channel into the price margin, a reward that will be phased out as the product becomes truly established in the mainstream, and competition for the right to distribute it increases.*

## Recap: Invasion Launching

To sum up, the last step in the D-day strategy for crossing the chasm is launching the invasion—that is, putting a price on your product and putting it into a sales channel. Neither of these actions resolves itself readily into a checklist of activities, but there are four key principles to guide us:

1. The prime goal is to secure access to a customer-oriented distribution channel. This is the channel you predict that mainstream pragmatist customers would want and expect to buy your product from.

2. The type of channel you select for long-term servicing of the market is a function of the price point of the product. If this is not direct sales, however, then during the transition period of crossing the chasm, you may need to adopt a supplementary or even an alternative channel—one oriented toward demand creation—to stimulate early acceptance in the mainstream.

3. Price in the mainstream market carries a message, one that can make your product easier—or harder—to sell. Since the only acceptable message is one of market leadership, your price needs to convey that, which makes it a function of the pricing of comparable products in your identified competitive set.

4. Finally, you must remember that margins are the channel's reward. Since crossing the chasm puts extra pressure on the channel, and since you are often trying to leverage the equity the channel has in its existing relationships with pragmatist customers, you should pay a premium margin to the channel during the chasm period.

This list of principles not only concludes this chapter but ties together chapters 3 through 7 on marketing strategy for crossing the chasm. The goal of these chapters has been to lay out a framework of marketing ideas to assist companies in meeting the challenges of the chasm period. The D-day strategy, as a whole, seeks to emphasize both the great peril and the great opportunity that lie before a company in this situation. The greatest impediment to action in such situations is often a lack of understanding of the appropriate alternatives. Hopefully, these chapters have gone some distance toward removing that impediment.

Having said all that, there is, finally, a larger set of issues that come into play. For if the chasm is a great challenge—and it is—it is one that is in large part self-imposed. To put it simply, our industry makes the chasm worse than it has to be. Until we understand how we do so, and stop doing so, we will never really master the chasm.

With this thought in mind, let us turn to our conclusion, "Leaving the Chasm Behind."

# Conclusion

# Leaving the Chasm Behind

---

It has become very fashionable of late to talk about how high-tech companies can and should become market-driven organizations. My own view, however, is that there is not any *becoming* involved. All organizations *are* market-driven, whether they acknowledge it or not. The chasm phenomenon—the rapid acceleration in market development followed by a dramatic lull, occuring whenever a discontinuous innovation is introduced—drives all emerging high-tech enterprises to a point of crisis where they must leave the relative safety of their established early market and go out in search of a new home in the mainstream. These forces are inexorable—they *will* drive the company. The key question is whether the management can become aware of the changes in time to leverage the opportunities such awareness confers.

Thus far we have been treating the chasm as a market development problem and have focused exclusively on marketing strategies and tactics for crossing it. But the impact of the chasm extends beyond the marketing organization to every other aspect of the high-tech enterprise. In this final chapter, therefore, we are going to step back from the marketing view and look at three other critical arenas of change: finance, organiza-

193

tional development, and R&D. The goal of the discussion in every case is the same—to keep the enterprise moving forward into the mainstream marketplace and not, as so often happens, to allow it to fall back into the chasm.

The fundamental lesson of this chapter is a simple one: *The postchasm enterprise is bound by the commitments made by the prechasm enterprise.* These prechasm commitments, made in haste during the flurry of just trying to get a foothold in an early market, are all too frequently simply unmaintainable in the new situation. That is, they promise a level of performance or reward that, if delivered, would simply destroy the enterprise. This means that all too often one of the first tasks of the postchasm era is to manage our way out of the contradictions imposed by prechasm agreements. This, in turn, can involve a major devaluation of the assets of the enterprise, significant demotions for people who are unsuited to the responsibilities implied by their titles, and marked changes in authority over the future of the product and technology—all of which is likely to end in bitter disappointments and deep-seeded resentment. In short, it can be a very nasty period indeed.

The first and best solution to this class of problems is to avoid them altogether—that is, *to avoid making the wrong kind of commitments during the prechasm period.* By looking ahead at the outset, while we are still in the early market phase, to where we must go in order to survive the chasm crisis, we can vaccinate ourselves against making the kind of crippling decisions that doom so many otherwise promising high-tech enterprises.

Let me acknowledge that this is much harder to achieve than it looks. I am reminded of the many times as an adolescent when I was sagely advised that I was in the process of making some very bad decisions because I was "going through a phase." I loathed that advice. First, it made me feel vaguely inadequate and rather inferior to the person giving it. And second, even though I suspected it to be true, it was totally useless information. I might be going through a phase, but since I was in the phase, and was therefore doomed to perform in some incompetent way, what good was this knowledge? How could I stop being myself?

That, however, is exactly what the high-tech enterprise must accomplish to leave the chasm behind. The enterprise must stop "being itself"—in the sense that it must accept that it is going through a phase and act competently with that knowledge.

To leave the chasm behind, there is a molting process that must occur, a change of company self, wherein we grow away from celebrating familial feelings and dashing individual performances and step toward rewarding predictable, orchestrated group dynamics. It is not a time to cease innovation or to sacrifice creativity. But there is a call to redirect that energy toward the concerns of a pragmatist's value system instead of a visionary's. It is not a time to forego friendships and implement an authoritarian management regime. Indeed, management style is one of the few things that can remain constant during this period of transition. But there is a call to review and revalue the skills and instincts and talents that helped to build up a leadership in the early market in light of the new challenge of building leadership in the mainstream. And that call can and will test friendships and egos throughout the firm.

The principles and practices for successful postchasm management of financial, organizational, and product development issues are all significantly different from their prechasm counterparts, and not everyone is adaptable or amenable to the changes required to operate in the new order. The good news is, in either case, there will always be plenty of jobs. That is, while individual high-tech enterprises have shown a very erratic track record over the past 10 years, the sum total of revenue and employment of the industry as a whole has grown dramatically. We all need to remember this during the chasm reshuffling. It should not be our goal, that is, to try to evangelize a new style of behavior but rather to create a framework for helping individuals understand for themselves where they best fit in, and then take appropriate action.

With that thought in mind, let us turn to the first and most influential set of decisions that postchasm enterprises inherit from their prechasm selves—the financial ones.

## Financial Decisions: Breaking the Hockey Stick

The purpose of the postchasm enterprise is *to make money*. This is a much more radical statement than it appears. To begin with, we need to recognize that this is not the purpose of the prechasm organization. In the case of building an early market, the fundamental return on investment is the conversion of an amalgam of technology, services, and ideas into a replicable, manufacturable product and the proving out that there is some customer demand for this product. Early market revenues are the first measure of this demand, but they are typically not— nor are they expected to be—a source of profit. As a result, the early market organization is not required to adopt the discipline of profitability.

Nor does the prechasm organization motivate itself by profitability, or typically any other financial goal. Oh, to be sure, there are the get-rich dreams that float in and out of idle conversation. But there are much headier rewards closer at hand—the freedom to be your own boss and chart your own course, the chance to explore the leading edge of some new technology, the career-opening opportunity to take on far more responsibility than any established organization would ever grant. These are what really drive early market organizations to work such long hours for such modest rewards—the dream of getting rich on equity is only an excuse, something to hold out to your family and friends as a rationale for all this otherwise crazy behavior.

So early market entrepreneurs are not called to focus on, nor are they oriented toward, making money. This has enormous significance, as most management theory assumes a profit motive present, serving as a corrective check against otherwise alluring tactics. When that motive is not present, people make financial commitments that have consequences they either do not, or do not care to, foresee. Although this comes in many and varied forms, perhaps its most prevalent one is the *hockey stick forecast of revenue growth*.

In the current, flawed model of high-tech market development—the two-stage one without the chasm—the entrepreneur is asked to drive the enterprise to an early market success and then to hand over the reins to professional managers who will guide the company as its revenues and profits skyrocket toward market leadership. This is the model traditionally endorsed by the venture capital community, the one it uses to attract its capital funds, and the one it applies to its investment opportunities.

If you do not show this kind of meteoric rate of return sooner or later, you are not qualified to participate in their portfolio.

Entrepreneurs may be many things when it comes to financial issues, but they are typically not slow on the uptake. If venture capitalists are the ones with the money, and these are the rules you follow to get that money, then they will be sure to follow the rules. And so entrepreneurs raise capital using "hockey stick" graphs of revenue attainment. That is, they bring forward a business plan that shows no revenue development for some period of time—as long as they possibly can defer—after which there is a sharp inflection in the curve, and rapid, continuous, and what any sane person would call miraculous, revenue growth from there on. As a form, it is as precise and conventional as a love sonnet—and just as likely to get one into trouble.

Hockey stick curves are created by spreadsheets, a software tool that many have argued has driven some of the worst of the investment decisions of the 1980s. It is so easy to increment a revenue number by a percentage and just let the software take it from there. Now in theory, this revenue line approximates a real profile of how the company could capitalize on a developing market opportunity. As such, it would serve as the "master line" in the spreadsheet, the one to which all others must account. That is how profitable operations work.

In fact, however, the revenue line is a slave—and to not just one but two masters. At the front end, it is slave to the entrepreneur's cost curve, and at the back, to the venture capitalist's hockey stick expectations. Revenue numbers, under this methodology, are . . . well, whatever they have to be. Once that sum is identified, then market analyst reports are scoured for some appropriate citations, and any other source of evidence or credibility is enlisted, to justify what is a fundamentally arbitrary and unjustifiable projection of revenue growth.

Now, if the current model of high-tech market development were not flawed, this might work, or at least work better or more often. But in fact, the revenue development that actually occurs looks more like a *staircase* than a hockey stick. That is, there is an initial period of rapid revenue growth, representing the development of the early market, followed by a period of slow to no growth (the chasm period), followed by a second phase of rapid growth, representing return on one's initial mainstream market development. This staircase can continue indefinitely, with the flat periods representing slower growth due to transitioning into broader and broader mainstream segments, and the rapid rises representing the ability to capitalize on those efforts. As more and more segments are served, sooner or later the ups and downs begin to cancel each other out, and one can achieve the less bumpy results that Wall Street greatly prefers. (In fact, only the most successful high-tech companies have achieved such a state; most continue to fluctuate more dramatically than the financial community can understand, with the result that their stocks routinely take a vicious beating at the slightest indication of bad news.)

All this is well and good. The staircase model is perfectly viable—unless you have mortgaged your stake in the company on making the hockey stick scenario come true. That, unfortunately, is precisely what most high-tech funding plans commit to. And when the hockey stick scenario does not come true, and the mortgage comes due, the founder's equity gets radically diluted, things fall apart, and the company dies in the chasm. That is the course sketched out in the high-tech parable in Chapter 1 of this book.

Now, the venture community has long been aware of this problem. Cynics in high tech believe they count on it—that's how the "vulture capitalists" take over the company from the unwitting entrepreneur. But the truth is, such a strategy is a lose/lose proposition, and most investors know it. They may call it "the valley of death" instead of the chasm, but they know it is there. All they have to do is look at their own portfolios.

The question now becomes, If we have the chasm model to work with, what can we do differently? This question really breaks into two parts—one directed to the financial communities that provide the sources of capital, and the other to the high-tech executives who provide the sources of management. For the former, the key issue is how to reformulate its concepts of valuation and expected rate of return, and for the latter, it is when to spend capital and when to adopt the discipline of profitability. Let's look at both of these more closely.

# The Role of the Venture-Financing Community

All investment is a bet on performance against competition within time. What the chasm model surfaces is a need to rethink these variables. From the investment point of view, the most pressing question initially is, How wide is the chasm? Or, to put this in investment terms, How long will it take before I can achieve a reasonably predictable ROI from an acceptably large mainstream market?

The simple answer to this question is, as long as it takes to create and install a sustainable whole product. The chasm model asserts that no mainstream market can occur until the whole product is in place. A reasonable corollary, I believe, is that once the whole product is in place—in other words, has become institutionalized—the market will develop quickly— normally, although not necessarily, around the company that drove and led the whole product effort.

Can we predict how long this will take? I think so. By analyzing the target customer and the compelling reason to buy, and then dissecting all the components of the whole product, we can reduce this process to a manageable set of performance factors, each of which can be projected ahead in time, with an estimated point of convergence. It's not a science, but it's not a black art either:  it is, in essence, just another kind of business plan.

Supposing this plan has some credibility, a raft of other questions immediately follow. How big will this market be? Again, the simple answer is, As big as can be motivated by the value proposition—the compelling reason to buy—and served by the whole product. Market boundaries occur, in other words, at the point of failure of either the value proposition or the whole product. The other market-making factors—alliances, competition, positioning, distribution, and pricing—do not impact the size of market but rather the rate of market penetration. Given free market economy incentives, efficient solutions in these areas will fall into place sooner or later if the market is truly there.

If all of the preceding assertions are true—and that is certainly something that warrants further investigation—then all the key factors of the investment decision are reasonably out in the open, and the decision itself can be made without having to consult the entrails of a sacrificial animal. Estimates of market size, rate of penetration, cost to achieve market leadership, and anticipated market share can all be made in the light of day, without smoke and without mirrors. There will still be plenty of room for disagreement about probability of success and degree of risk, but there should not be any fundamental leap of faith demanded, no "drinking of the Kool-Aid" as one of my more macabre colleagues has put it.

So the call to action to the investment community is, Make your client companies incorporate crossing the chasm into their business planning. Demand to see not only broad, long-term market characterizations but also specific target customers for the D-day attack. Drive them to refine their value propositions until they are truly compelling, and then use these to test how many target customers there truly are. Force them to define the whole product, and then help them to build relationships with the right partners and allies. Again, use the results to test hypotheses about market size. As for competitive set and positioning, beware of pushing your small fishes too soon into big ponds. And as for distribution and pricing, don't look for "standard margins" until the chasm has truly been crossed. To sum up, use the crossing-the-chasm matrix of ideas to ensure proper management of financial assets.

## The Role of the Venture-Managing Community

Now let's turn to the entrepreneur's key concern: How long should I live off of capital, and when should I adopt the discipline of profitability? The bounds of this decision work as follows. Until profitability is achieved, nothing is secure, and your destiny is not under your own control. This argues for early adoption. In fact, in slow-developing markets, particularly in the software industry, which has low capitalization requirements, there is a very strong case for adopting profitability from day one. Early visionary customers will pay consulting fees and prepay royalties to help fund low capitalization start-ups. From an accounting view, these prepaid royalties cannot be booked immediately as revenue, but they can make you cash-flow positive from day one, and thus keep 100 percent of the equity reserved for a later date.

The great benefit of adopting the discipline of profitability at the outset is that you do not have to learn it later on. All too frequently, even when they are led by experienced managers, enterprises that are funded for long periods of time fall into a "welfare state mentality," losing their sense of urgency, and looking for their next paycheck to come from yet another round of financing instead of from the marketplace. Moreover, the discipline of profitability teaches you to "just say no" early and often. For most ideas there simply isn't any money to fund them. The enterprise is forced to focus drastically just because of resource constraints. This radically reduces time to market because people are not focused on doing something else and because they understand it is the market that is paying their paychecks. And finally, when one does go seeking external capital, there is no stronger evidence for a high company valuation than it having already demonstrated not only real market demand but its own ability to process that demand profitably.

Indeed, the case for seeking profitability from the beginning is so strong, you begin to wonder why you would ever not choose this route. Essentially, there are two reasons. First, the price of entry is too great to fund with sweat equity or consulting contracts. This is clearly the case in any manufacturing-intensive operation. Entry into the pen-based laptop market, according to the executives at Momenta with whom I am currently working, is a $50 million proposition, and Intel estimates it can cost ten times that much to build a new semiconductor fabrication plant. This kind of investment is not likely to come out of working capital. Indeed, there is some question whether it can come out of the United States—but that is a topic for another book.

The other reason to forego initial profitability is when the market is expected to develop so rapidly that you cannot afford to mark time as a bit player. The recent rollout of Microsoft Windows 3.0, for example, has created a new window of opportunity for software applications in the PC industry. It is not necessarily clear that the current DOS market leaders can transfer their hegemonies into the new environment. Indeed, Microsoft's own applications group is betting that they cannot. The point is, there is a very limited time during which there will be a level playing field, and it is critical for vendors to have their products in the market during this period. For once the new leaders are chosen, then the market will return to the conspiracy behavior that keeps them in place.

Beyond this there is a third, more general principle that can help entrepreneurs think through their management of capital. It is typically more capital intensive to cross the chasm than it is to build the early market. Early market development efforts typically do not respond well to massive infusions of capital—the IBM PC Jr. being an old case in point, and NeXT, I would argue, being a new one. You simply cannot spend your way into the hearts and minds of technology enthusiasts and visionaries. To be sure, there is a minimum level of capitalization required. You have to be able to travel to make direct sales calls, and show up looking presentable, and you probably should have an office and a phone that is answered in a professional way. You do need to invest in early market public relations—the product launch is crucial to building early market success—but you do not need to advertise, nor do you need to invest in developing partnerships or building channel relationships. All this is premature until you have established some early market credibility on your own.

Once early market leadership has been established, however, the entire equation changes. The whole product investment—securing the partnerships and alliances and then making them work to deliver the final goods—takes a significant number of funded initiatives. So does the channel development process, both on the pull and on the push sides, creating demand and providing incentives for sales. And it is critical during this period to have an effective communications program, including press relations, market relations, and advertising.

In sum, this is when you want to spend your money—not before. It is important, therefore, that you not start this process until after you have established early market leadership, and that you not commit to throwing off all kinds of cash during the chasm period. Simply applying these two concepts to the business plan can keep you out of a lot of trouble.

## Organizational Decisions:
## From Pioneers to Settlers

Turning from issues of finance to issues of people, we must recognize that the chasm separates not only visionaries from pragmatists—it also separates the companies that serve them. To leave the chasm behind, to cross it and not fall back into it, involves a transformation in the enterprise that few individuals can span. *It is the move from being pioneers to becoming settlers.*

In the development organization, pioneers are the ones who push the edge of the technology application envelope. They do not institutionalize. They do not like to create infrastructure. They don't even like to document. They want to do great deeds, and when there are no more great deeds to be done, they want to move on. Their brilliance fuels the early market, and without them, there would be no such thing as high tech.

Nonetheless, once you have crossed the chasm, these people can become a potential liability. Their fundamental interest is to innovate, not administrate. Things like industry standards and common interfaces and adaptations to installed solutions, even when these solutions are clearly technically inferior, are all foreign and repugnant to the high-tech pioneers. So as the market infrastructure begins to close in around them, they are already looking for less crowded country. In the meantime, they are not likely to cooperate in the compromises needed, and can be highly disruptive to groups that are seeking to carry this agenda out. It is critical, therefore, that as the enterprise shifts from the product-centric world of the early market to the market-centric world of the mainstream, that pioneer technologists be transferred elsewhere—ideally, into another project within the enterprise, but if necessary, to another company.

There is a comparable process going on in the sales force at the same time. Here the group at the forefront is the high-tech sales pioneers. These are people who have the gift of selling to visionaries. They are able to understand the technology and product at a level where they can readily manipulate it and adapt it to the dreams of the visionaries. They can talk the visionaries' language, understand the quantum leap forward visionaries seek to achieve, and wrap their products in that

cloak. They can translate that language back into concrete manifestations of the product, to be illustrated through custom
demos, for which they make insatiable demands. They can
think big, and they can get big orders. They are the darlings of
the early market. Without them, achieving early market leadership is all but impossible.

These same people, however, also become a liability once you
have crossed the chasm. Indeed, they are the ones primarily
responsible for dragging companies back into the chasm. The
problem is, they cannot stop making the visionary sale, a sale
predicated on delivering custom implementations of the whole
product. Such contracts are fulfilled by robbing from Peter—the
mainstream R&D effort—to pay Paul—the custom R&D effort
necessary to achieve the visionaries' buying objective. The key
to leaving the chasm behind, however, is to stop custom developments and institutionalize the whole product, to build to a
set of standards that the marketplace as a whole can support.
This mainstream effort necessarily puts enormous strain on the
R&D department, who must not, therefore, be distracted by yet
another wild and crazy venture. And so it is that a pioneer
salesperson left unchecked can be highly disruptive and demor-
alizing to a sales organization looking to leave the chasm
behind.

So now we have two sets of people—high-tech pioneers and
pioneer salespeople—who are fundamental to success in the
early market and potentially a liability after the company has
crossed the chasm. They must be outplaced, but who is competent to do so? And how in the world will their knowledge ever
be replaced? And who is going to take over what they leave
behind? And is any of this moral or fair, given their contributions to date?

I know of no high-tech firm that has not struggled with these
issues sooner or later. And how you respond affects not only
those who leave but those who stay. This is a time when you
must perform impeccably.

Let's deal with the moral issue first. And let us take as our
starting point that casting aside people, dislocating their lives
and threatening their livelihood, is immoral—even if businesses
and governments routinely do so with abandon. The issues
then become ones of foresight, agreement, planning, and prepa-

ration. Pioneers do not want to settle down. That is not in their best interest nor in the interest of the companies that employ them. If, at the beginning of the process, everyone can acknowledge this fact, and acknowledge that the very goal of pioneers, the final manifestation of their success, is to create a mainstream market and thereby put themselves out of a job, then we can have a reasonable basis for going forward. How we would go forward and under what kind of compensation program is a discussion we need to postpone until we look at how to make the transition to the other side of the equation, to the settlers who are expected to come in and take their place.

The truth is, of course, that settlers do not take pioneers' places. They take other places, ones that pioneers never have occupied nor would ever choose to. Nonetheless, settlers do take over the employment roster, and the management positions and the authority and, ultimately, the budget. And they build fences and create laws (called procedures) and do all the things that created range wars between pioneers and settlers back in the Old West. All this bodes well for the postchasm marketplace, populated with pragmatists, who like reliable, predictable people and abhor surprises. But it hardly sits well with the pioneers. How in the world, then, can you make the transition between these two groups in an orderly way?

## Two New Job Descriptions

The key is to initiate the transition by introducing two new roles during the crossing-the-chasm effort. The first of these might be called the *target market segment manager*, and the second, the *whole product manager*. Both are temporary, transitional positions, with each being a stepping stone to a more traditional role. Specifically, the former leads to being an industry marketing manager, and the latter to a product marketing manager. These are their "real titles," the ones under which they are hired, the ones that are most appropriate for their business cards. But during the chasm transition they should be assigned unique, one-time-only responsibilities, and while they are in that mode, we will use their "interim" titles.

*The target market segment manager has one goal in his or her short job life—to transform a visionary customer relationship into a potential beachhead for entry into the mainstream vertical market that particular customer participates in.* If Citicorp is the client, then it is banking; if Aetna, insurance; if Dupont, chemicals; if Intel, semiconductors. The process works like this.

Once you have closed such an account as part of an early market sales program, assign the target market segment manager as its account manager with a charter that allows him the kind of extensive customer contact that will let him really learn how their business works. He must attend the trade shows, read the literature, study the systems, and meet the people— first, just within the one account, and subsequently, in related companies. At the same time, he must take over the supervision of the visionary's project, make sure it gets broken up into achievable phases, supervise the introduction and rollout of the early phases, get feedback and buy-in from the end users of the system, and work with the in-house staff to spin off the kind of localized implementations that give these initial deliverables immediate value and impact. At the same time, he will be working with the whole product manager to identify which parts of the visionary project are suitable for an ongoing role in the whole product and which are not. The goal is to isolate the idiosyncratic elements as account-specific modifications, making sure thereby not to saddle the ongoing product development team with the burden of maintaining them.

The market segment manager should not be expected to generate additional revenue from the account in the short term, because the visionaries believe they have already paid for every possible modification they might need. What he can be expected to do, however, is the following:

- *Expedite the implementation of the first installation of the system.* This not only contributes to the bottom line, as it will expedite the purchase of additional systems, but it also secures the beginning of a reference base in the target market segment. Most companies fail miserably in this regard, so much so that even several years later their initial "big name" accounts cannot be referenced. The key here is to remember that pragmatists are not interested in hearing about who you have sold to but rather who has a fully implemented system.

- *During the implementation of the first installation, introduce into the account his own replacement, a true account manager, a "settler," who will serve this client, hopefully, for many years to come.* Note that at this point the pioneer salesperson is still in the picture, still has the relationship with the visionary, but that the day-to-day operation of the account is entirely in others' hands. This is typically just fine with the pioneer, for he recognizes this to be the kind of detail-oriented settler work for which he has no liking.

- *Leverage the ongoing project to create one or more whole product extensions that solve some industrywide problem in an elegant way.* The intent is either to absorb these elements into the product line or to distribute them informally as an unsupported product extension through a users' group. Either way, such add-ons increase the value of the product within the target market segment and create a barrier to entry for any other vendor.

## The Whole Product Manager

While the target market segment manager is pursuing these tasks in the customer's environment, there is a corresponding internal role to be filled. Here the transition is from product manager to product marketing manager via the short-lived role of whole product manager. These titles are all sufficiently alike as to be confusing, so let's take a minute to sort out these three very different jobs.

A *product manager* is a member of either the marketing organization or the development organization who is responsible for ensuring that a product gets created, tested, and shipped on schedule and meeting specification. It is a highly internally focused job, bridging the marketing and development organizations, and requiring a high degree of technical competence and project management experience.

A *product marketing manager* is always a member of the marketing organization, never of the development group, and is responsible for bringing the product to the marketplace and to the distribution organization. This includes all of the elements on the crossing-the-chasm agenda, from target-customer identification through to pricing. It is a highly externally focused job.

Not all organizations separate product managers from product marketing managers, but they should. Combining the jobs almost always results in one or the other simply not getting done. And the type of people who are good at one are rarely good at the other.

Now, the *whole product manager* is a product-marketing-manager-to-be. The reason she is not one today is that the job itself is premature. Until there is a successful crossing of the chasm, there are no meaningful market relationships or understandings to drive the future of product development. The target market segment manager is off getting these under way, but they are not there today. What is there today, on the other hand, is a list of bug reports and product-enhancement requests that is growing with disconcerting speed. *If this list is not managed properly, it will bring the entire development organization to its knees.*

The tactic, which at once secures proper management of the list and initiates a transition process from pioneer to settler culture in the development side of the house, is to take this list away from the product manager and give it to the whole product manager. For whoever is serving as the product manager at this point almost certainly is a pioneer—otherwise, the organization could not have got to where it is today. The problem with this person continuing to direct the future of the product is that she will be driven first and foremost by her own personal commitments made to early customers. Unfortunately, these commitments are often not in the best interest of the mainstream market customer. To be sure, they must eventually be fulfilled—unless they are to be negotiated away—but in either case, they should not be given automatic priority over other issues. What should increasingly become the prioritizing factor for ongoing product development work is contribution to mainstream, pragmatist customer satisfaction—in other words, contribution to the whole product—hence, the need to transfer authority.

Once this authority is transferred, the enterprise has taken a key step in moving from a product-driven to a market-driven organization. As the shape of the mainstream market emerges, as the needs of this market can be increasingly identified through market research and customer interviews, then the whole product manager steps into the title that she has had all along on her business card, product marketing manager. To try to take this step earlier in the market development cycle is foolish. During the early market it is important to be product-driven and to give strong powers to the product manager. But to fail to take it now is equally foolish, for every day that the enhancement list is in the hands of the original pioneers, the company risks making additional development commitments to unstrategic ends.

To sum up, at the beginning of the chasm period, the organization is dominated by pioneers, with strong powers invested in a few top-gun salespeople and product managers. By the time we are into the mainstream market, that power should be distributed far more broadly among major account managers, industry marketing managers, and product marketing managers. This gradual dissemination of authority will ultimately frustrate the pioneer contributors, hampering their ability to make quick decisions and rapid responses. Ultimately, it will make them want to leave.

## Coping with Compensation

This brings us back, full circle, to the fundamental issue that underlies so much of the frustration and disappointment that builds up within high-tech organizations—compensation. Few compensation programs recognize either the fundamentally different contributions of pioneers and settlers or their fundamentally different tenures with the enterprise, and thus these programs end up discriminating against one or the other. And when compensation programs do discriminate—when they discourage the very behaviors that ought to be rewarded, or vice versa—then organizations fail.

To work through all the complexities of designing appropriate compensation schemes is beyond both the scope of this book and the capabilities of its author. I can only sketch out a few general principles that seem important to follow.

First, let's start on the sales side. A typical pioneer sale involves a broad purchase agreement, predicated on successful implementation of a pilot project. Even when there has been a major up-front payment, the rational way to book this business is to defer recognizing the larger order until it has been confirmed. That could be at least a year away, and during that period, we will have introduced a number of new players into the account, including the target market segment manager. The pioneer salesperson might even be gone by then. Say, some account manager just joins the firm, inherits the account, and all of a sudden the flood of orders come in. What is the appropriate way to compensate?

*The key is to discriminate between account penetration and account development. The latter is a more predictable, less remarkable achievement. It is also the more lucrative.* Compensation here should reward such things as longevity of the relationship, customer satisfaction, and predictability of revenue stream. It should be spread out over time and not clumped into dramatic payments. Because there is high value associated with the intangibles of the ongoing customer relationship, much of it can be based on an MBO formula rather than pure revenue attainment. If equity is part of the compensation strategy for the firm as a whole, it is a reasonable component here was well, provided it is doled out slowly, with the larger portions coming at the end of the program, to reward stability of service. Overall, however, since this is not a high-risk role, it should not be a high-reward one either.

*Compensation for the pioneer salesperson should have the opposite characteristics. It should provide the bulk of its rewards immediately, in recognition of a single key achievement—winning the account.* This is an extraordinary event, one that few can accomplish, and it is critical to determining the firm's long-term future. It is an extraordinarily high risk endeavor, with the odds stacked heavily against the salesperson. It therefore deserves extraordinary compensation. On the other hand, if it was achieved by promising more than anyone can deliver, perhaps even more than anyone really knew, then that is not behavior we want to reward. So, although we would like the compensation to be front-loaded, there must also be a reality check built into the process. Because the pioneer salesperson will be moving on, we do not want an extended compensation program, and thus equity, for example, is an inappropriate vehicle. Taking all this together, the situation argues for a bonus-based program more than a straight commission approach—something lucrative for the salesperson, event-driven and over and done with relatively quickly, and not so closely tied to revenue recognition that either the pioneer has to overstay his or her welcome in order to reap the rewards or earns an extraordinary cash reward at a time when the company simply cannot afford that sort of out-lay.

## Compensating Developers

Moving over to the development side, there is one remaining compensation challenge—the pioneer technologist. These divide into two camps—true company founders and very early employees. The former have bet their lives on the equity gam-ble, and there is nothing further to discuss, except to hope that in reading this book they learn to conserve a large portion of that equity to fund crossing the chasm. The latter pose a real problem. They can point with accuracy to the notion that they created a large part of the core product. Thus, should that prod-uct become a mainstream market hit, they feel they should get a major share of the gains. The fact is, they don't, and the truth is, bluntly, they don't deserve it either. Mainstream success, as we have argued at length, is a function of the whole product, not the core product, and that is a very large team effort indeed.

What the pioneer technologist does have a right to is a large share of the early market returns, because here it truly is the core product that drives success. The problem is that cash is typically so tight during this period that there is none to throw off in the form of a reward. So equity is the usual fallback. This is a compromise, to say the least, as equity should be reserved for people who cross the chasm and stay—not the pioneer's ideal role.

The final word on pioneer technologists, I suppose, is that they are in the same bind as authors—a fate I can identify with. Like authors, they are compelled to conduct their craft regard-less of whether anyone will pay for it. As such, their negotiating position is fundamentally weak, and their normal compensa-tion reflects it.

To sum up, improper compensation wastes dollars and demotivates people. To be appropriate to high tech, compensa-tion programs must take into account the differences between desired performance in the early market and mainstream market, as well as the types of people that can be called on to achieve these performances, and the likelihood that some of these people will need to leave the company long before it achieves significant profitability. If we can sort through these issues, and come up with an appropriate distribution of rewards, we can forego much of the agony and loss of momentum that accompany most crossings of the chasm. If we continue to operate the way we do today, we will persist in constructing self-conflicting organizations and wonder why they are not more productive.

## R&D Decisions:
## From Products to Whole Products

At the outset of this book, we set crossing the chasm as the fundamental marketing priority in high tech. In the middle we established that institutionalizing the whole product was the fundamental strategy for succeeding in this endeavor. It is fitting, therefore, to finish up with a look at the impact of whole product marketing on long-term R&D.

R&D is high tech. Everything else is secondary. As an industrial sector, before anything else, we are technology-driven. Eventually we learn to create products, and then markets, and then enterprises to dominate those markets. But it starts with technology. "Build the product and they will come," to paraphrase the theme of the movie *Field of Dreams*. That is our fundamental dream, the dynamic that drives all else.

The problem is, we grow past the dream. The products and markets and companies we create all grow up to make persistent and legitimate demands on us, and we have no choice but to serve them. *And once this scenario begins, R&D doesn't get to focus on the generic product any more. It must become whole product R&D.*

Whole product R&D is driven not by the laboratory but by the marketplace. It begins not with creative technology but with creative market segmentation. It penetrates not into protons and processes but rather into habits and behaviors. It does not, like the captain of the starship Enterprise, "go where no man has gone before," but rather, like T. S. Eliot, finds the end of all its exploring is "to arrive where we started and know the place for the first time." It prefers to assemble its creations from existing technologies and products rather than to invent new ones from scratch. Its heroes are less like Einstein, who developed a whole universe out of his own head, and more like George Washington Carver, who discovered over three hundred different uses for the peanut.

Not very heady stuff. No wonder it is so often ignored. Indeed, the word that high tech uses for whole product R&D is *maintenance*. And the people they assign to it are . . . well, the janitorial types. No top guns want to go near this stuff.

Instead, the top guns rush out to create more discontinuous innovations, flooding the market with far more technology that it can possibly absorb, and complaining all the while about how product life cycles are becoming shorter and shorter. They play the game, in other words, almost entirely to the left of the chasm, cycling though endless repetitions of early markets that never cross over to the mainstream. *Product* life cycles truly are getting shorter—but *whole product* life cycles are as long as they ever were. Ask Lotus about 1-2-3. Ask Microsoft about their current best-selling product (no, not Windows—Works). What is Apple's biggest seller as we enter 1991? Not their hottest new top-of-the-line Mac II but the Mac Classic. There's gold in them thar hills.

## An Emerging Discipline

Whole product R&D is an emergent discipline. It represents a kind of convergence between high-tech marketing and consumer marketing, where, for the first time, the tools of the latter can be of significant use in solving the problems of the former. Let's look at two examples: focus groups and packaging studies.

As innovation becomes increasingly continuous, focus groups, which are virtually useless in guiding the development of an early market, become effective tools. The reason they are now effective is that the fundamental product proposition is already in the market and absorbed. Until this is the case, consumers are way over their head in trying to anticipate the value and usage of a new high-tech product. But once that proposition is in place, the tool becomes effective. Specifically, it can be used to direct the extension and modification of an existing product line to meet the special needs of a target market segment. In this context, all consumers are asked to do is address relatively minor derivatives from a known entity—something well within their expertise. The information they give back, therefore, is valuable.

Consider another discipline that today is far more advanced in consumer marketing than in high tech—packaging. As an industry, we have considered this to be nothing more than the paint of the box, the logo, the cover. But packaging happens not just on the outside but on the inside, and the goal of good packaging is to ensure a successful experience right out of the box— an area that cries out for more research attention in high tech. Think how many dollars could be diverted into better ends that today go to expensive support services, all because our products are packaged in confusing or obtuse ways.

Now these types of efforts—focus groups and packaging studies—are traditionally located in the marketing department. But in high tech, marketing is too ignorant to drive the bus. What appears to the generalist to be a simple change may in fact cut across some fundamental technology boundary in a radically inappropriate way. Or conversely, what looks impossible to achieve may in fact be a by-product of a minor adjustment. In either case, engineering must be a direct partner in the effort, or it is wasted. It's not market research alone, nor is it just product development. It's whole product R&D, and it implies a new kind of cooperation between organizations traditionally set apart from each other.

## Leaving This Book Behind

By way of parting, let us look back over the ground we have covered in this and the previous chapters. We began by isolating a fundamental flaw in the prevailing high-tech marketing model—the notion that rapid mainstream market growth could follow continuously on the heels of early market success. By analyzing the characteristics of visionaries and pragmatists, we were able to see that a far more normal development would be a chasm period of little to no growth. This period was identified as perilous indeed, giving companies every incentive to pass through it as rapidly as possible.

Taking such rapid passage as our charter, we then embarked on setting forth strategy and tactics for accomplishing it. The fundamental strategic principle was to launch a D-day type of invasion, one focused on a highly specific target segment within a mainstream marketplace. The tactics for implementing that invasion were then set out in four clusters.

To begin with, we had to *target the point of attack*, which meant isolating our target customers and their compelling reason to buy. Then we had to *assemble the invasion force*, constructed around the whole product and the partners and allies needed to make it a reality. The next step was to *define the battle*, by creating our competition and positioning ourselves, in that context, as being easy to buy. Finally, we had to *launch the invasion*, selecting our intended distribution channel and setting our pricing to give us motivational leverage over that channel.

Now we have just spent this last chapter stepping back from the immediate tactics of crossing the chasm, to look at the major commitments that get made in the prechasm phase of an organization's growth, thereby to guard against crippling the success of the postchasm venture. That brings us to the end of this road.

Finally, it should come as no surprise that there are no warranties, expressed or implied, on any of the methods described in this book. You must use them at your own risk. But I do claim that they are the best I know of, and that they are representative of "best practices" as conducted at Regis McKenna Inc. On behalf of my colleagues, as well as myself, I wish you the best of success in all your upcoming marketing efforts.

# Index